# POISONOUS PLANTS OF PENNSYLVANIA

**Robert J. Hill**
**Botanist**

**Donna Folland**
**Illustrator**



**Pennsylvania Department of Agriculture**
**Bureau of Plant Industry**
**2301 N. Cameron Street**
**Harrisburg, PA 17110-9408**



**1986**

ISBN 0-8182-0078-2

# CONTENTS

# INTRODUCTION

From the dawn of time plants have played a major role in the human drama: in our work, our recreation, and our struggle for life. They have a place in our myths, religions, and histories, as well as in the arts, medicine, and economics. Plants are mostly beneficial — there are actually few poisonous types of plants considering the thousands encountered in the home or natural surroundings. Pennsylvania has perhaps a hundred toxic species; it is advantageous to be aware of these poisonous plants in our environment.

Each year the Department of Agriculture, physicians, poison control centers, hospitals, veterinary clinics, and local high school and college biology teachers are contacted about actual or suspected plant poisonings. During the 4 year period (most recent data) from 1978-1981, the U.S. National Clearinghouse for Poison Control Centers reported almost 50,000 cases for all age categories in which the ingested substance was plant material. Of this number, slightly more than 10% (5,500) were classified as toxic, i.e. cases reported with signs, symptoms, hospitalization, or death. Two of the three deaths for this period were listed as suicide. The fatal agents were jimsonweed and mayapple. The overall low mortality rate is attributable, at least in part, to accurate identification of the toxic plant, and rapid response by persons attending the victims. It is hoped that this volume contributes to the aid of the poisoned victim.

The number of cases of toxicosis in livestock far outweighs those reported for humans. No accurate statistics are available, but it is estimated that as many as several thousand animals, in all classes of livestock, die annually in the U.S. from plant toxicosis.

Therefore, it is with pleasure that I introduce the reader to *Poisonous Plants of Pennsylvania.* This publication is designed to offer a diverse audience the most current information on the poisonous plants encountered in the Commonwealth. It is an enlarged, revised edition that builds on the merits of previous works. The history of such works is not insubstantial. In 1935 the State Botanist, E.M. Gress, published an authoritative booklet on this subject. Reprinted in 1953, it remained the principal reference until succeeded by a new edition in 1965, by Wendell P. Ditmer. The year 1935 also witnessed the printing of a poisonous plants booklet by Edward Graham, published by Carnegie Museum, Pittsburgh. Also, approximately one hundred poisonous plants from western Pennsylvania were listed in a mimeographed circular by L.K. Darbaker (1940).

I am greatly indebted to the authors of several invaluable publications on poisonous plants. Having familiarized myself with these publications, I freely drew from the wealth of information contained therein. Their laborious and painstaking contribution is acknowledged. They include the works of Kingsbury (1964), Hardin and Arena (1974), Lewis (1977), Kinghorn (1979), and Scimeca and Oehme (1985).

The excellent work of Dale Wallace, who typed the manuscript and its revisions, is appreciated. Dr. A. G. Wheeler, Jr. gave invaluable assistance in editing the manuscript. His time, patience, and expertise are gratefully acknowledged.

Importantly, this work was begun as a cooperative program between the Pennsylvania State University and the Pennsylvania Department of Agriculture. Bruce Young, an intern to the Bureau of Plant Industry, laid the foundation during the 1984 summer. He accompanied the author on poisonous plant calls, began collecting information, performed library researches for references, and collated existing data. His tireless assistance is acknowledged in gratitude.

Finally, this book would not have been possible without the constant encouragement and assistance I received from Donna Folland. The artistic quality and scientific accuracy of the illustrations stand as a testimony to her ability and scholarship. The tedious task of final editing and technical details in the publication of this work are credited to her, and thankfully acknowledged.

# HOW TO USE THIS BOOK

This volume is intended for use by homeowners, farmers, veterinarians, and health care professionals, as well as botanically curious individuals. Each poisonous plant is listed alphabetically by scientific name. Scientific and common names and illustrations are listed in the index. Because it was necessary to use some technical terminology in this publication, a glossary of botanical terms is provided. The metric system is used throughout. For those unfamiliar with this system, a Conversion Table is provided on the back cover.

**PLANT NAMES:** Each entry provides the generic name, followed by the Latin scientific and common names of the species that are poisonous in the genus. When all species in a genus are known or suspected to produce toxicosis, the abbreviation "spp." is listed after the generic name.

**FAMILY:** The family to which each plant belongs is fully described. Occasionally a family will contain several genera of poisonous plants. To avoid duplication of family characteristics, the reader is referred to the genus where the family description first appears.

**PHENOLOGY:** Phenology is the study of flowering time of a plant in relationship to climate. This may be a useful characteristic in determining the proper identity of a species.

**DISTRIBUTION:** Distribution data are provided. The habitat where a poisonous plant is found can be helpful. It should be noted that some toxic plants are ornamentals, others are houseplants, and still others are native or naturalized species.

**PLANT CHARACTERISTICS:** Both the genus and species have diagnostic features necessary to properly identify the toxic plant. This section, like the family description, will often contain terminology defined in the glossary.

**POISONOUS PARTS:** This is a listing of the plant organs (flowers, leaves, seeds, roots, etc.) that contain the poisonous substances. It should be noted in some instances that not all parts of a plant are equally toxic. Toxicity and lethal dosages of the plant material, when available, are noted here.

**SYMPTOMS:** Symptoms are listed for each poisonous plant. In some instances postmortem lesions, both gross and histological, are provided.

**POISONOUS PRINCIPLES:** This section details the toxic substances produced by the plant. Natural toxic substances can be alkaloids (water insoluble, bitter tasting principles), amines and polypeptides (molecules used to build proteins), glycosides (several classes exist: cyanogenic, goiterogenic, coumarin, steroid and triterpenoid, cardiac, and saponins), mineral toxicosis agents (nitrogen), oxalates (corrosive oxalic acid), photosensitization agents (pigments or liver-damaging substances), phytotoxins (protein molecules), resins and resinoids (complex compounds not necessarily related).

**CONFUSED TAXA:** This category provides information necessary to differentiate poisonous plants from plants with which they might be readily confused.

**TREATMENT:** This section enumerates antidotes or therapeutic care that can be administered to aid the recovery of a poison victim. Treatment categories often overlap for different cases of toxicosis. To avoid repetition, therapies are listed by number. Each number is fully explained in Appendix I at the end of the book.

**OF INTEREST:** Finally, pertinent facts and items, fascinating or otherwise illuminating, are listed.

Most entries are accompanied by line illustration. The drawings show diagnostic features necessary for accurate identification. The majority of illustrations were executed by the artist from living material. If such material was unavailable, pressed herbarium specimens were used, and the drawing labeled 'ex herbario', a Latin phrase meaning "from the herbarium".

# POISONOUS PLANTS OF PENNSYLVANIA

# Acer rubrum

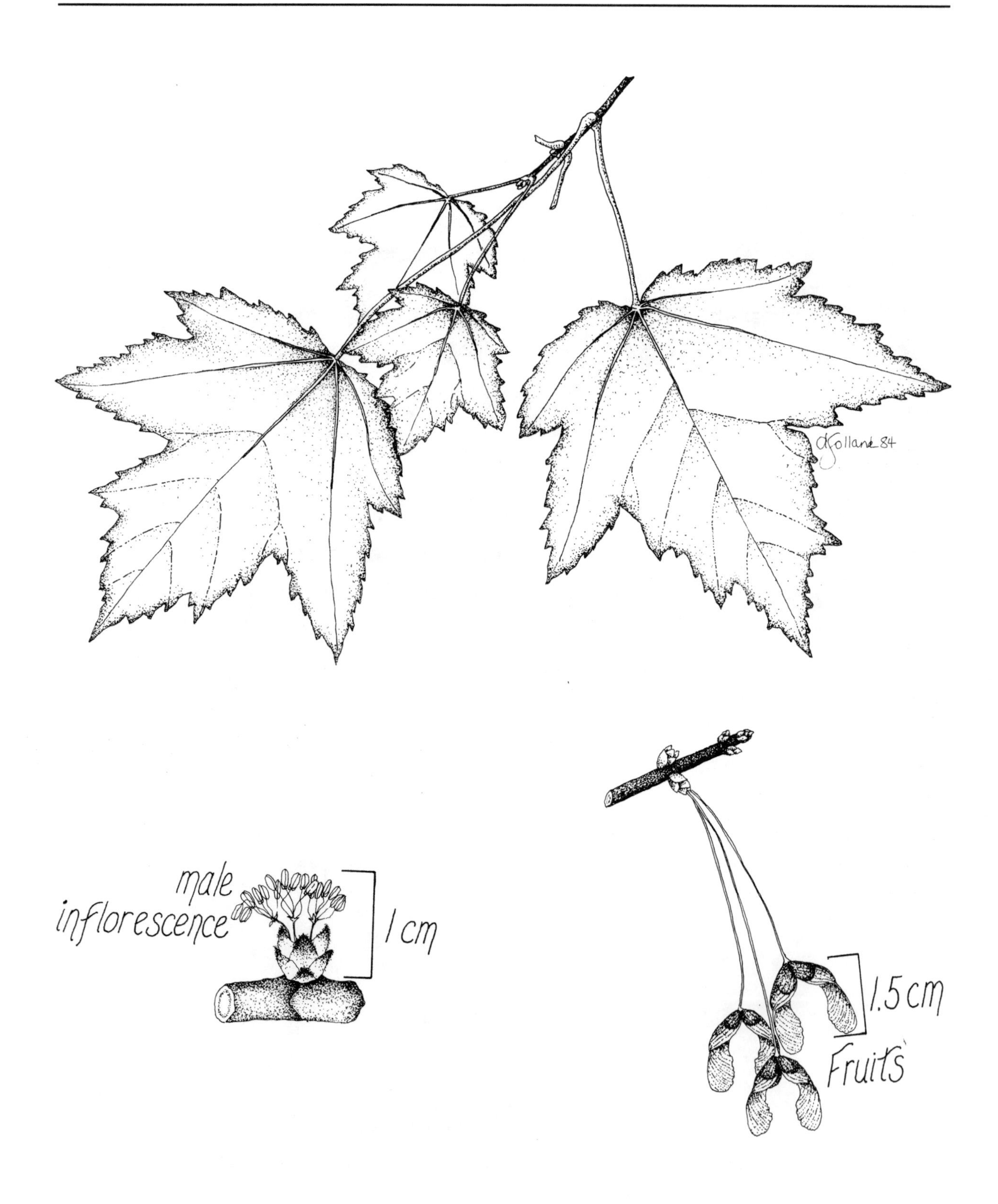

# GENUS: *Acer*

*Acer rubrum* L. — Red maple

**FAMILY:** Aceraceae — the Maple Family

The maples are generally well known. **Flowers** are completely or functionally unisexual, usually 5-merous; **petals:** small, separate, or lacking; **stamens:** often 8 (5 to 10); **ovary:** superior, 2-celled, producing a pair of winged, 1-seeded fruits; **trees** or **shrubs** with opposite, simple, or occasionally compound **leaves.**

**PHENOLOGY:** *Acer rubrum* flowers March through May; fruits mature May through June.

**DISTRIBUTION:** Found in swamps, moist uplands, and on alluvial soil.

**PLANT CHARACTERISTICS:** **Tree** to 35 m tall; **leaves** sharply but shallowly lobed, coarsely double-serrate or with a few minor lobes.

**POISONOUS PARTS:** The leaves are responsible for livestock poisoning. Apparently only wilted leaves are toxic, with toxicity remaining in the leaves for about a month.

**SYMPTOMS:** Within 18 to 24 hours after consumption of red maple leaves, horses begin to show yellow or brown discoloration of the mucous membranes, especially gums and eyelids, urine becomes dark red to brown, and animals become febrile (102.0-103.5°). About 50% of the horses that consume red maple leaves are affected. As many as 64% of those affected die, usually from methemoglobinemia, a destruction of hemoglobin in the blood.

**POISONOUS PRINCIPLES:** The toxic agent(s) is unknown.

**CONFUSED TAXA:** There are 10 common species of maples in eastern United States. *Acer rubrum* is the only maple with all of the following characteristics: simple leaves; angled, sharp sinuses between the principal leaf lobes; flowers appearing much before the opening of the leaf buds; and glabrous fruit maturing in the spring.

**SPECIES OF ANIMALS AFFECTED:** Only horses and ponies have been reported to develop toxicosis from eating red maple leaves.

**TREATMENT:** The best treatment is prevention. Do not graze animals in areas where red maples occur. Do not pile branches or leaves in places where stock can reach them. When removing a red maple tree from an area frequented by horses, do it in the winter when the leaves are absent.

# Actaea

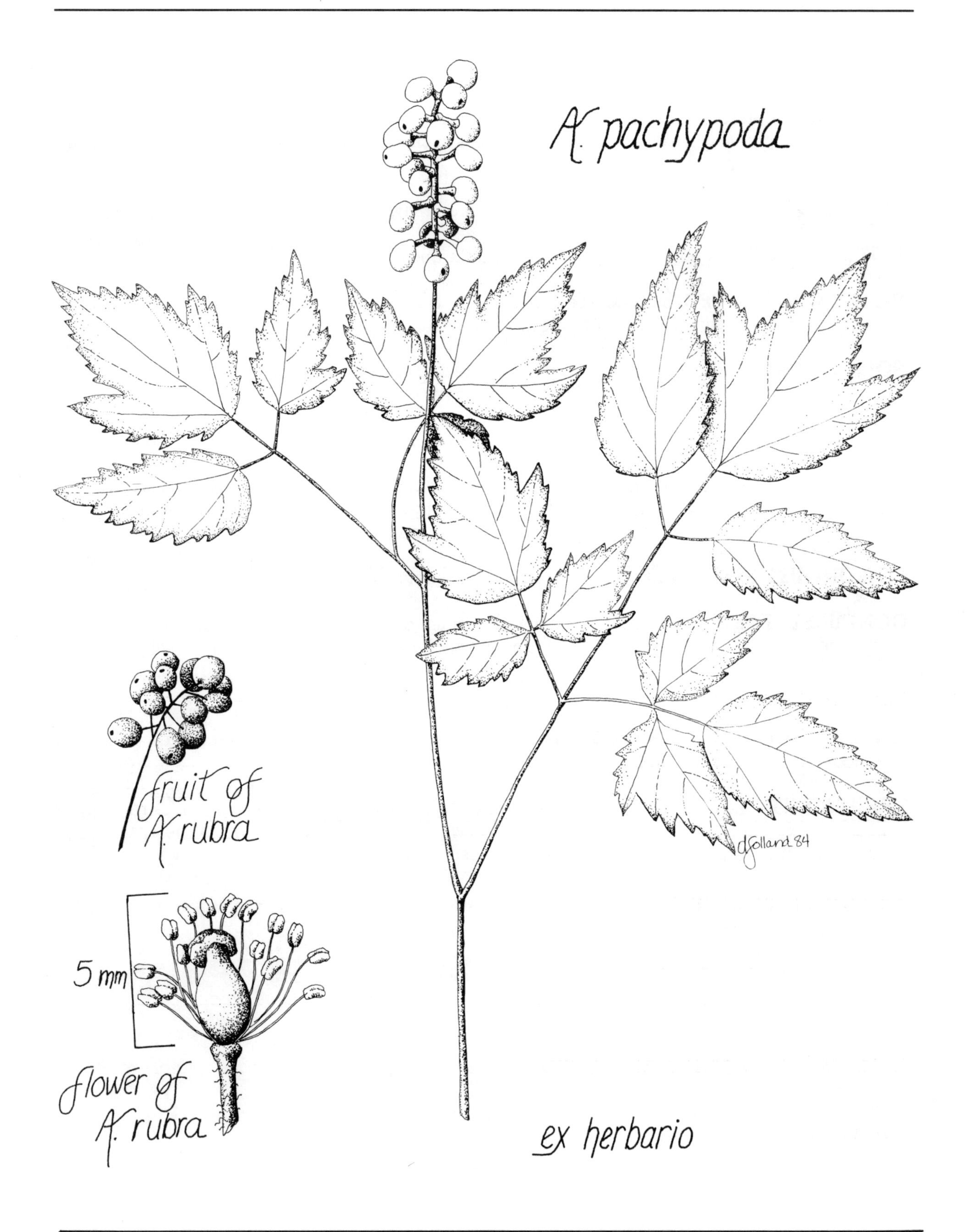
A. pachypoda
fruit of
A. rubra
5 mm
flower of
A. rubra
ex herbario
dJolland 84

# Genus: *Actaea*

*Actaea pachypoda* Ell. — Baneberry; dolls-eyes
*Actaea rubra* (Ait.) Willd. — Baneberry; dolls-eyes

---

**Family:** Ranunculaceae - the Buttercup or Crowfoot Family

This is a large family of plants containing many genera, including several that produce acrid-narcotic poisons. The family is so diverse that only a general description is provided. The **plants** are predominately herbaceous, with colorless, acrid juice; **sepals:** 2 to many; **petals:** numerous or in some species absent, with the calyx colored like the corolla; **stamens:** rarely few, typically very numerous; **pistils:** few to many and spirally arranged (1 in *Actaea*); **fruits:** dry capsules, seedlike achenes, or berries; sepals, petals, stamens, and pistils all distinct and unconnected; **leaves:** often dissected, petioles dilated at the base, sometimes with stipulelike appendages. Many genera are cultivated for ornamental purposes, some contain medicinal properties, and one, *Nigella,* produces edible seeds used as an herb. In some genera the sepals or petals are saccate, producing **spurs** that often function as nectar-holding organs.

**PHENOLOGY:** *Actaea* species flower through May and June.

**DISTRIBUTION:** *Actaea* is found in moist, rich woods. *A. pachypoda* occurs throughout the Commonwealth, whereas *A. rubra* is more often encountered in the northern half of the state.

**PLANT CHARACTERISTICS:** *Actaea* has 3 to 5 petaloid, caducous **sepals; petals:** 4-10, deciduous, clawed; **stamens:** numerous, filaments elongated and widened upward; **pistil:** 1; **stigma:** broad, sessile, 2 lobed; **fruit:** a several-seeded berry; **perennial herbs** to 1 meter tall from a thick rhizome; **leaf blades:** large, (2-) 3-ternately compound; **leaflets:** with sharply toothed margins; **flowers:** small, white in dense, long-stalked terminal racemes. *Actaea pachypoda* has white **fruit** (rarely red), the stigma wider than the ovary, and very stout fruiting pedicels. *Actaea rubra* has red **fruit** (rarely white), the stigma more narrow than the ovary, and slender fruiting pedicels.

**POISONOUS PARTS:** All parts, especially roots and berries, are toxic. As few as six berries have been reported to cause severe symptoms.

**SYMPTOMS:** Conditions include gastroenteritis with associated acute stomach cramps, dizziness, vomiting, increased pulse, delirium, circulatory failure, and headache. Symptoms usually disappear after 3 hours. Losses of life in the United States have not been reported for these plants; however, the European literature chronicles deaths of children after eating berries of a European species of baneberry.

**POISONOUS PRINCIPLES:** The toxic compound is unknown but probably is an essential oil or poisonous glycoside.

**CONFUSED TAXA:** The (2-)3-ternately compound leaves resemble many species of forest plant. However, the conspicuous fruits (either white or red) terminated by a black "button" (aging stigma) are characteristic of *Actaea*.

**SPECIES OF ANIMALS AFFECTED:** Both livestock and humans are susceptible to *Actaea* toxins.

**TREATMENT:** (11a)(b); (26)

# Aesculus

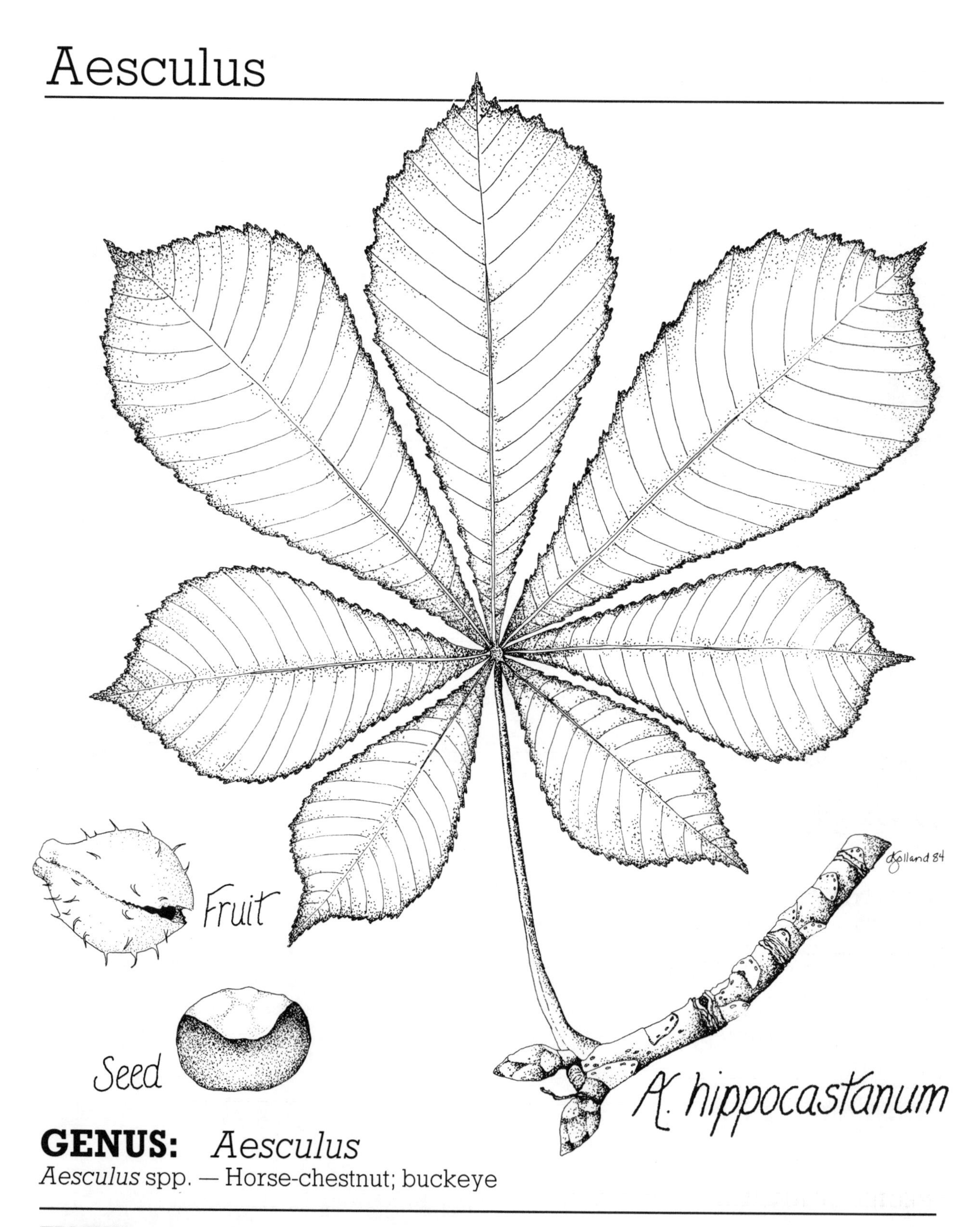


## GENUS: *Aesculus*

*Aesculus* spp. — Horse-chestnut; buckeye

**FAMILY:** Hippocastanaceae — the Horse-chestnut Family

This family consists of **trees** or **shrubs** with opposite, palmately compound **leaves** composed of 5 to 7 serrated leaflets. **Flowers** are zygomorphic, perigynous with an extra staminal, often 1-sided disk; **sepals:** 5, united at least half their length; **petals:** 4 or 5, white, yellow, or red, clawed; **stamens:** 5-8, with elongated and often exserting filaments; **ovary:** 3-celled, 2 ovules in each chamber; **style:** elongate; **fruit:** a leathery, globose capsule bearing sharp spines when young, becoming smooth in some species, eventually opening by 3 valves; **seeds:** 1 (subglobose), 2 (hemiglobose), or 3 (flattened sides), glossy brown, bearing a large conspicuous, light-brown scar, usually about 2.5 cm in diameter.

**PHENOLOGY:** Depending on the species, the flowering period can be May to June.

**DISTRIBUTION:** Some species are cultivated; of these a few occasionally escape and become established. Other taxa occur naturally on moist, alluvial soil, in rich moist woods, or along streams.

**PLANT CHARACTERISTICS:** *Aesculus hippocastanum* L. is the common horse-chestnut. Widely cultivated, it is a tree to 25 m with variously colored, double-flowered and hybrid forms available. Older stock has the white upper and lateral petals marked with red or yellow at the base; petals number 5. *Aesculus glabra* Willd. (Ohio buckeye) is a small tree with yellow flowers found predominantly along river banks and in moist woods in western Pennsylvania and now introduced in the eastern part of the state; it has 4 greenish-yellow petals, long-exserted stamens, and spined fruit; *Aesculus octandra* Marsh. (sweet buckeye) has 4 yellow (sometimes purple or red) petals, stamens barely exserted, and smooth fruit. Cultivated species of *Aesculus* include the dwarf or bottle brush buckeye (*A. parviflora* Walt.) with white flowers and long-exserted stamens and the red horse-chestnut (*A. carnea*), a hybrid tree from *A. hippocastanum* x *A. pavia.*

**POISONOUS PARTS:** Nuts (seeds), stump sprouts, bark, flowers, leaves, dried fruits, and young growth are dangerous.

**SYMPTOMS:** Experimental feeding of *A. pavia* L. produced the following symptoms on ingestion at the rate of 1% of the animal's weight: incoordination, nervous twitching of muscles, sluggishness, and excitability. It is reported that flowers are poisonous to honey bees; poisoning of humans eating honey produced from the nectar of California buckeye has been reported. In Europe ingestion of the seeds has reportedly killed children, and leaves and dried fruits have caused loss of cattle. Other symptoms may include dilated pupils, vomiting, diarrhea, depression, paralysis, and stupor.

**POISONOUS PRINCIPLES:** Alkaloids, glycosides, and saponins are responsible for toxicosis. One important constituent is aesculin (esculin), a lactone glycoside and hydroxy derivative of coumarin. This molecule shows a chemical relationship with the toxic substance in spoiled sweet-clover hay (*Melilotus* spp.), which also contains coumarin glycosides. Sweet-clover poisoning is a hemorrhagic disease fatal to cattle. Loss is due to internal or external hemorrhage.

**CONFUSED TAXA:** The erect, large, many-flowered inflorescences may superficially resemble those of catalpa trees (*Catalpa* spp.), which have large, simple, heart-shaped leaves, and the Princess Paulownia tree (*Paulownia tomentosa* (Thunb.) Steud.), which also has large, simple leaves. *Aesculus* fruits may be confused with those of several species of trees. American beech (*Fagus grandifolia* Ehrh.) has small, sharply 3-angled fruits. They are 1-seeded and normally born in pairs within an accrescent, 4 valved involucre. The prickles of the fruit are 4-10 mm, erect to spreading or recurved and numerous. Another genus, *Castanea,* the chestnut, bears solitary, 2-or 3-seeded fruits. These are enclosed within an accrescent, long-spined, 2-to 4- valved involucre; the stiff spines are numerous, more than 10 mm long, and often branched from the base.

**SPECIES OF ANIMALS AFFECTED:** All classes of livestock and humans are potential victims of *Aesculus* poisoning upon consumption of this plant.

**TREATMENT:** (11a)(b); (26)

**OF INTEREST:** *Aesculus* was a source of medicinal preparations in past years. The common name horse-chestnut is derived from the belief that Turks fed a kind of "chestnut" to their horses to enable them to breathe more easily.

# Agrostemma

# Genus: *Agrostemma*

*Agrostemma Githago* L. — Corncockle

---

**FAMILY:** Caryophyllaceae — the Pink Family

This family is an economically important group because of the large number of ornamental plants it contains. These include the florist's carnation, baby's-breath, maltese cross, sandworts, pinks, and others. Only corncockle *(Agrostemma Githago)* and bouncing Bet (see *Saponaria*) are poisonous; both are common in Pennsylvania. Characteristics of the family include: opposite **leaves; petals:** distinct, 5 (sometimes none); **sepals:** separate or connate, 5; **stamens:** 1-10, commonly twice as many as the petals; **ovary:** superior, 1- to 3-celled, mostly 1-celled with **ovules** (and seeds) attached to a central, basal column that is not fused to the top of the ovary; **stigmas** and **styles:** 2-5; **fruit:** a capsule, opening at the apex by valves or teeth of the same number (or twice the number of the styles); **stem:** often swollen at the nodes.

**PHENOLOGY:** Corncockle has an extended flowering period, July through September.

**DISTRIBUTION:** *Agrostemma Githago* is widely established as a weed of grainfields and waste places. Infrequently it is cultivated as a garden plant. The seeds are difficult to separate from wheat seeds and may contaminate this product.

**PLANT CHARACTERISTICS: Stems:** often 1 m, thinly hairy to silverish; **leaves:** 8-12 cm x 5-10 mm, without petiole or stipules, linear or lanceolate; **flowers:** red, conspicuous; on pedicels to 2 dm, solitary at the ends of branches; **calyx:** fused, 12-18 mm; 5 calyx lobes 2-4 cm long; **petals:** 5, each 2-3 cm long, notched at the apex; **styles:** 5; **stamens:** 10; **fruit:** 14-18 mm, the capsule bearing numerous black seeds; **seeds:** covered with small warts and pits.

**POISONOUS PARTS:** The seeds are primarily responsible for poisonings from corncockle; however, all parts are suspected to be toxic. Seeds consumed at a concentration of 0.2-0.5% of body weight are lethal to young poultry; older birds are less susceptible.

**SYMPTOMS:** The toxic response includes severe gastroenteritis, acute stomach pain, vomiting, diarrhea, dizziness, listlessness, weakness, and slow breathing.

**POISONOUS PRINCIPLES:** The toxin is primarily the sapogenin githagenin, which may be 5- 7% of the weight of seeds.

**CONFUSED TAXA:** In our area few members of the Caryophyllaceae have large, red flowers. Some *Lychnis* species superficially resemble *Agrostemma,* but in *Lychnis* the petals are appendaged and the calyx lobes are much shorter than the tube. *Agrostemma* petals lack an appendage, and the calyx lobes are longer then the calyx tube, often surpassing the petals.

**SPECIES OF ANIMALS AFFECTED:** Poultry, horses, and other livestock are susceptible. In animals that vomit freely (e.g. pigs), acute poisoning is less likely.

**TREATMENT:** (11a)(b); (26)

**OF INTEREST:** Flour milled from wheat contaminated with corncockle has caused human poisonings. Current agricultural methods have largely eliminated this problem.

# Amanita

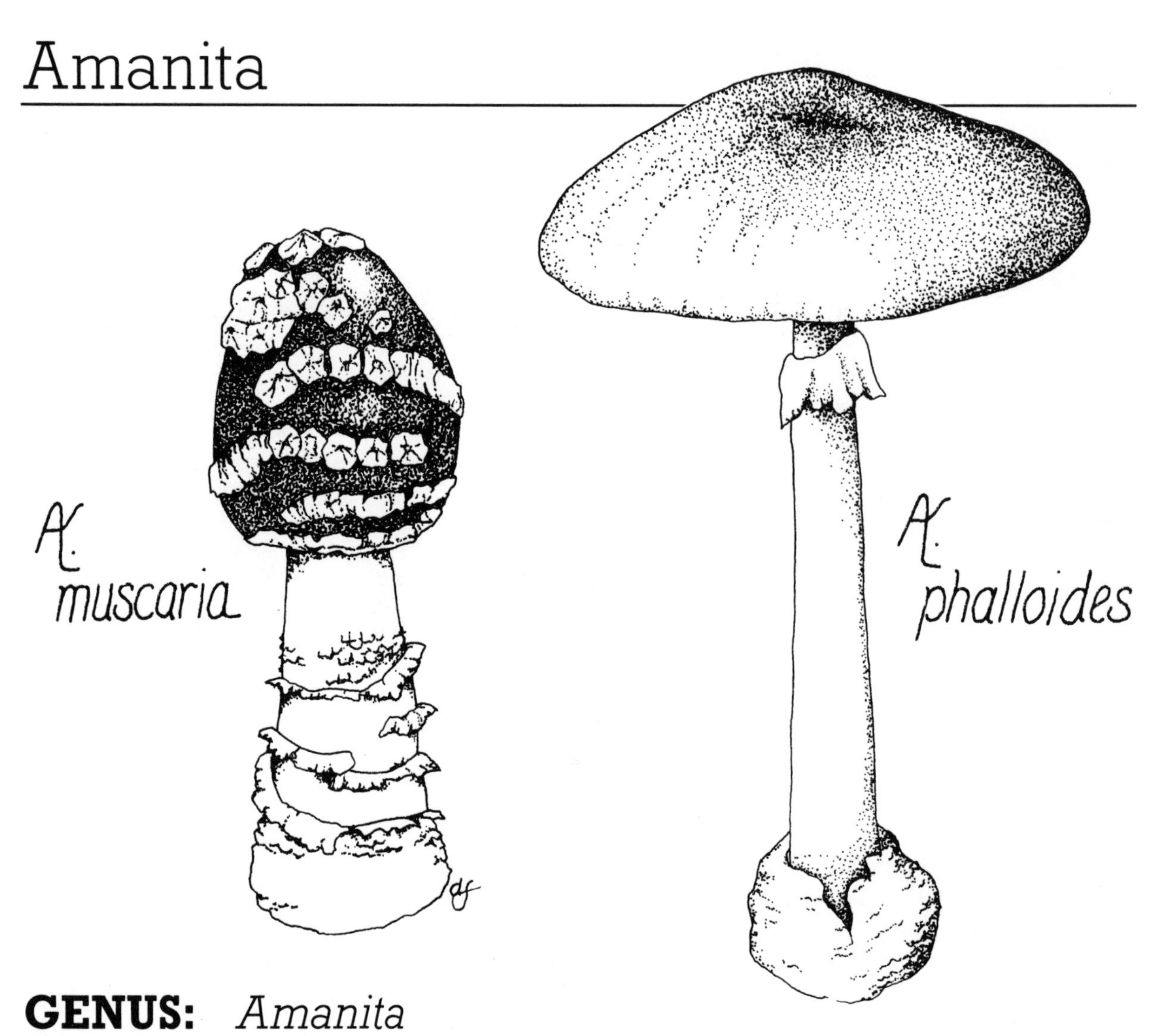


## GENUS: *Amanita*

*Amanita muscaria* (Fr.) S. F. Gray — Fly amanita; fly mushroom; fly agaric
*Amanita phalloides* Fries — Death cap

**FAMILY:** Amanitaceae — the Amanita Family

Amanitas begin as round or oval **buttons** covered by a protective layer, the **universal veil.** The young button mushroom has small and complete gills, cap, and stalk and can be mistakenly identified as edible puffballs, often with deadly results. As the stalk grows the universal veil is torn, appearing on the expanding cap as warts or patches of tissue. If the universal veil is thick or tough, it will be split by the growing cap and stalk. The cap is then devoid of remnants, but a well-formed cup, the **volva,** surrounds the base of the stalk. The volva occasionally remains in the soil when a specimen is collected; therefore, the absence of a basal cup on a specimen may be misleading when attempting to identify the amanitas. The **gills** are free from the stipe. **Spores** of the Amanitaceae are entire, smooth, and thin walled. A further microscopic feature is the divergent tissue in the center of the gill; it grows outward from a central strand.

**DISTRIBUTION:** The amanitas are found singly or in numbers under hardwoods and conifers from the spring through the fall.

**PLANT CHARACTERISTICS:** *A. muscaria:* **cap:** 8-24 cm across, convex or flat, bright yellow to orange red, surface rough with white or yellow wartlike spots; **gills** and **stem:** white; **stem:** 8-15 cm long and 20-30 mm thick; **base of stem:** bulbous; **veil:** white and persistent.

*A. phalloides:* This species is taxonomically complex, and occasionally several species are lumped under this name. The group includes *A. verna* (Bull.) Quel., *A. virosa* (Fr.) Quel., and *A. bisporiger* Atk. Recent evidence suggests that *A. phalloides* is rare and often confused with the more common *A. brunnescens,* which also is poisonous. True *A. phalloides* has a yellowish-green to green **cap** and white **veil** and **gills;** it is deadly poisonous. In *A. brunnescens* the **cap** is dark brown; in the deadly poisonous *A. virosa* the fruiting body is pure white and the **cap** is devoid of warts.

**POISONOUS PARTS:** All parts of the amanitas are poisonous.

**SYMPTOMS:** The characteristic, well-defined symptoms of *A. muscaria* poisoning may occur within 3 hours after ingestion. They include increased secretions from salivary, lacrymal, and other glands; perspiration; and possible severe gastroenteritis; much watery diarrhea plus retching and vomiting; possible labored breathing; pupils that are rarely responsive; and possible auditory or visual hallucinations or confusion occurring before or during the digestive upset. For *A. muscaria,* deaths are rare, but in such cases delirium is followed by convulsions, then coma with death from respiratory failure. In some severe cases, the patient may experience a profound sleep lasting a few hours, then awake without symptoms or memory of the illness that preceded.

Symptoms for the more deadly poisonous amanitas include a 10-hour lag period (6-15 hours) before onset of conditions. They begin as sudden, severe abdominal pain, vomiting, and diarrhea. Blood, mucus, and undigested food are present in vomitus and stool. Thirst, anuria, prostration, and restlessness are also present. If quantities of mushrooms are consumed, death ensues in 2 days; more typically the disease lasts 6 to 8 days before death in adults, 4 to 6 days in children. Fever, hematuria, tachycardia, hypotension, rapid volume depletion, and fluid and electrolyte imbalance also may be present.

**POISONOUS PRINCIPLES:** *A. muscaria:* The toxins are choline, muscarine, and muscaridine. The $LD_{50}$ i.v. in mice is 0.23 mg/kg. *A. phalloides* and other deadly amanitas contain amanitine and phalloidine (complex polypeptides). The toxins amanitin and amanin, also present, are highly toxic; the $LD_{50}$ i.p. in albino mice is 0.1 mg/kg; for phalloidine it is 3.3 mg/g i.m.

**CONFUSED TAXA:** Among the many species of *Amanita,* some are deadly poisonous, whereas others are nonpoisonous. Some of the deadly amanitas are rare (*A. phalloides*) or infrequent (*A. flavorubescens*); some are very common (*A. muscaria, A. virosa*).

**SPECIES OF ANIMALS AFFECTED:** Humans, all livestock, and wildlife are susceptible.

**TREATMENT:** *A. muscaria:* (11a - with 1:2,000 tannic acid or 1:10,000 potassium permanganate) or (11b); (5 - 0.1 to 0.5 mg either IM or IV, repeated as necessary). Atropine sulfate is antidotal.

*A. phalloides:* Mortality is 50-90%. First empty the stomach, then: (1-1 to 2 tablespoons in $H_2O$): corticosteroids and both peritoneal dialysis and hemodialysis to eliminate toxins and circumvent kidney failure. A high protein diet and intravenous doses of protein hydrolysate may prevent liver damage. Antiphalloidian serum is effective only when administered at the onset of symptoms; (26); thioctic acid, charcoal hemoperfusion, and vitamin C may be useful.

**OF INTEREST:** Mushroom poisoning can be produced by about 100 of the 2,000 species known. In the U.S. mushrooms of the genera *Amanita* and *Galerina* are the common causes of poisoning. Even trained mycologists may confuse toxic varieties with nonpoisonous or edible ones. There are no simple tests to identify poisonous mushrooms, no effective means to detoxify deadly kinds, and no simple rules or characteristics to follow in determining the toxicity of a mushroom.

Some inky cap mushrooms (*Coprinus* spp.) may produce toxic reactions if alcohol (beer, wine, etc.) is consumed with them. Some mushrooms (*Psilocybe* spp.) are hallucinogenic, contain psilocybin and psilocin, and are used in illegal drug trafficking.

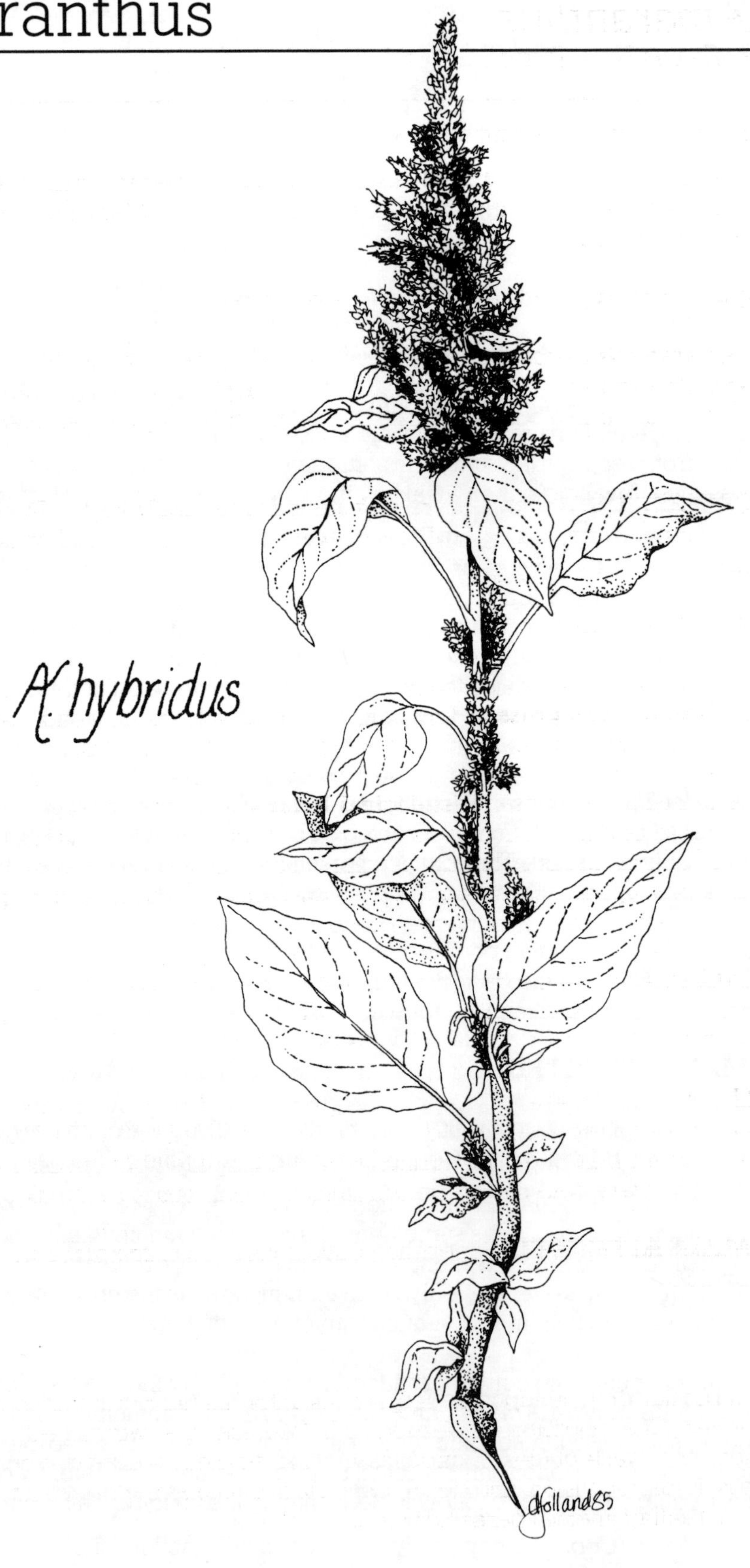
A. hybridus
Holland85

# GENUS: *Amaranthus*

*Amaranthus retroflexus* L. — Pigweed; redroot

**FAMILY:** Amaranthaceae — the Amaranth Family

The Amaranthaceae is a widely distributed family of herbs. **Flowers:** small, often unisexual, subtended by dry scales, frequently in showy cones; **fruit:** a utricle. Many species are weedy; some are grown as ornamentals.

**PHENOLOGY:** Pigweeds produce flowers in mid- to late summer.

**DISTRIBUTION:** *Amaranthus retroflexus* is a weed in Pennsylvania, as is the more common, closely related smooth pigweed (*A. hybridus* L.). The weedy amaranths, native to tropical America, are distributed in Pennsylvania in gardens, cultivated fields, pastures, roadsides, waste places, and fields.

**PLANT CHARACTERISTICS:** *Amaranthus retroflexus* is a tall **annual plant,** to 2 m; **leaves:** long-petioled, ovate or rhombic-ovate, to 1 dm; **inflorescence:** a terminal panicle of densely crowded spikes, 5-20 cm long.

**POISONOUS PARTS:** The foliage is poisonous.

**SYMPTOMS:** Losses have occurred to livestock, especially sheep. Cattle and horses are relatively resistant. **Postmortem: gross lesions:** perirenal edema (pigs, growing calves) possibly containing blood, affected kidneys pale yet normal in size; rectal and abdominal wall edema; distended pleural and peritoneal cavities caused by straw-colored fluid; kidneys may have ecchymotic hemorrhages in the cortex. **Histological lesions:** interstitial edema in renal cortex, tubular nephrosis, necrosis and dilation of the convoluted and collecting tubules with protein casts. Hydrothorax and hydroperitoneum is more pronounced in calves. In ruminants methemoglobinemia may appear. Blood and body tissue appear chocolate brown. Nitrite and ammonia ions may cause the stomach mucosal surface to be congested.

**POISONOUS PRINCIPLES:** The nephrotoxic agent(s) is not known. Oxalates, which are found infrequently as crystals in histological studies, may account for some symptoms. Toxic concentrations of nitrates also are responsible for toxicity.

**CONFUSED TAXA:** Among *Amaranthus* spp. occuring in the Commonwealth, the two most common are *A. retroflexus,* described above, and the related *A. hybridus.* The floral bracts are rigid, tapering to a point, 4-8 mm long in the former, whereas they are stout-tipped, 2-4 mm long in the latter. All species of *Amaranthus* should be considered dangerous to livestock.

**SPECIES OF ANIMALS AFFECTED:** Sheep, hogs, and young calves are more susceptible than adult cattle and horses.

**TREATMENT:** (26)

**OF INTEREST:** Lambsquarter (Chenopodiaceae) is a distantly related weed that contains at least one poisonous species, *Chenopodium ambrosioides* L. (Mexican tea, wormseed). It is found in gardens, roadsides, and wasteplaces. Symptoms include nausea, vomiting, abdominal pain, headache, dizziness, impaired vision, and depression. Oil of chenopodium contains ascaridol, an anthelmintic used in treating internal parasitic worms.

1 cm
1 cm
Holland 84
Veratrum viride
Liliaceae
ex herbario
Amianthium muscaetoxicum

# GENUS: *Amianthium*

*Amianthium muscaetoxicum* (Walt.) Gray - Fly poison

---

**FAMILY:** Liliaceae — the Lily Family

Found worldwide, this major group of plants contains predominantly perennial herbs having various **rootstocks:** rhizomes, bulbs, corms, or tubers. The family characteristics are; **flowers:** mostly bisexual, radially symmetric; **perianth:** usually large and showy, divided into 2 series called **tepals; sepals:** (outer whorl) may not be readily distinguished in shape or color from the **petals** (inner whorl); **stamens:** routinely 6; **pistil:** 1; **ovary:** superior, generally with 3 chambers (trilocular); **ovules:** distributed down the central axis. Economically the lily family is very important for its members used in horticulture. The cultivated ornamental plants include autumn crocus *(Colchicum),* tulip, Star-of-Bethlehem, hyacinth, lily, scillas, grape hyacinth, lily-of-the valley, and other bulbs of "Dutch trade." For a discussion of poisonous, cultivated members of the Liliaceae see *Colchicum, Convallaria,* and *Ornithogallum.* Crops in this family include onions (and related alliums) and asparagus. Garden plants include day lilies *(Hemerocallis)* and bishop's coat *(Hosta).* Wild flowers in the family include several major poisonous plants (see *Veratrum*) as well as minor elements. Many of these constitute a substantial portion of our "spring flora" such as trillium, true and false Solomon's seals, and dog-tooth violet (trout-lily).

**PHENOLOGY:** Fly poison flowers June and July.

**DISTRIBUTION:** *Amianthium muscaetoxicum* inhabits open woods and moist areas, often on acid soils.

**PLANT CHARACTERISTICS: Tepals:** 6, glandless, several-nerved, white to green, 1 cm wide; **stamens:** 6, filaments flattened; **ovary:** 3-lobed, deeply cleft and appearing like separate units, each lobe with a stout, conic style with minute stigma; **capsule:** containing 1 or 2 oblong, purple-brown seeds per cell; **perennial plants** growing from a thick **bulb** 5-8 cm in the ground; **basal leaves:** linear, 4 dm x 2 cm; **stem:** appearing later than basal leaves; **stem leaves:** much reduced; **racemes:** at first conic, becoming cylindric; **stalks** to 10 dm or more.

**POISONOUS PARTS:** Bulbs and leaves arc toxic.

**SYMPTOMS:** Symptoms include salivation, nausea, rapid or irregular breathing, staggering, weakness, lowered temperature, coma, and death due to respiratory failure.

**POISONOUS PRINCIPLES:** An unknown alkaloid(s) is the probable cause of toxicosis.

**CONFUSED TAXA:** Grasslike basal leaves and tall stalks with white flowers are not uncommon in the Liliaceae. Some of the confused taxa may be poisonous (e.g. *Melanthium* spp., which cause nervousness, anorexia, dyspnea, nausea, slobbering, sweating, weakness, stupor, weakened heart rate, and respiration), while others may not be. A botanical specialist should be consulted if *Amianthium* poisoning is suspected.

**SPECIES OF ANIMALS AFFECTED:** Experimental feedings have determined that sheep and cattle are susceptible (sheep death was produced from administration of leaves equal to 0.5% of the animal's weight). Losses may occur in early spring when little else is available for livestock forage.

**TREATMENT:** (11a)(b); (1)

# Anagallis

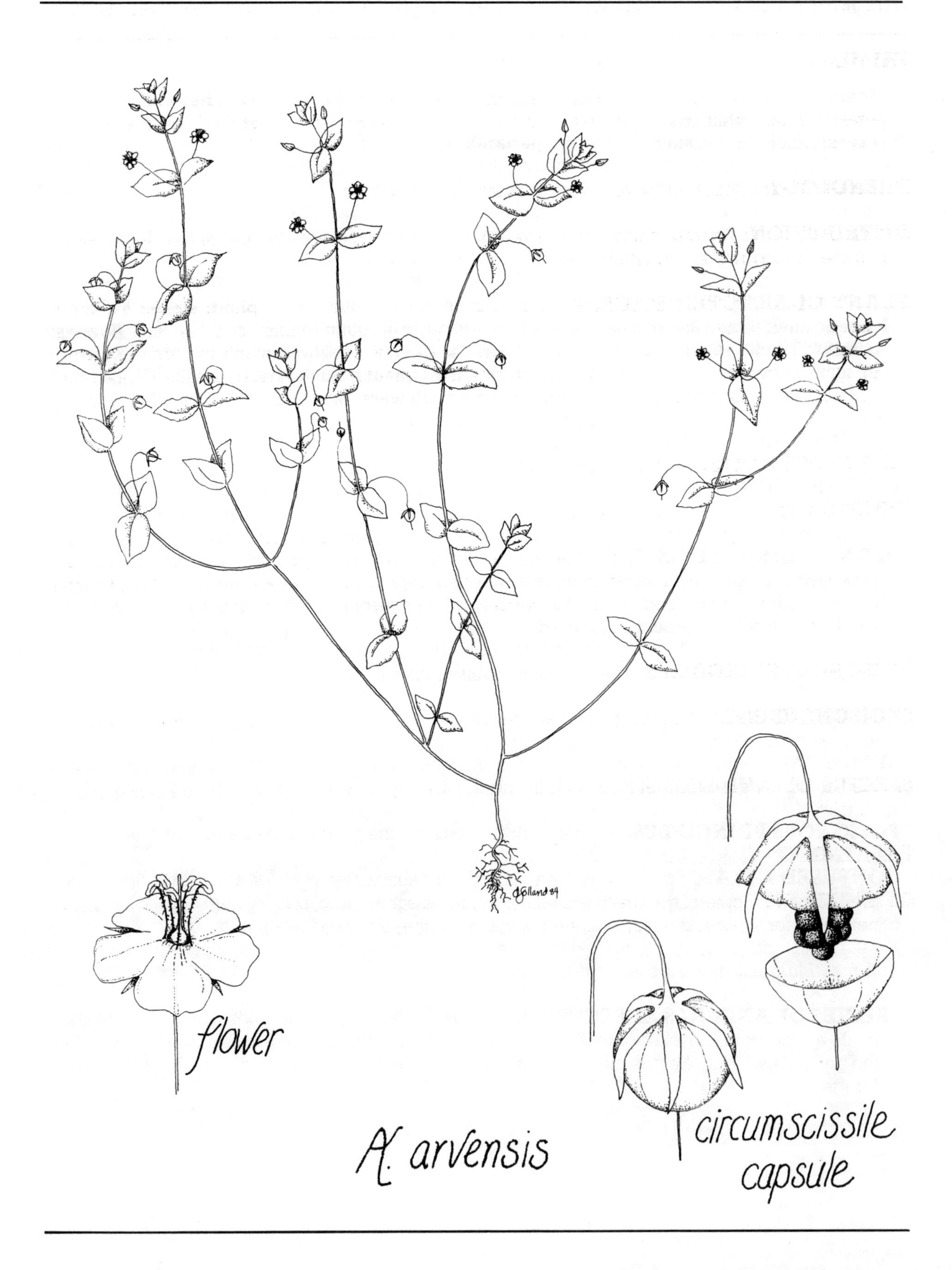

# GENUS: *Anagallis*

*Anagallis arvensis* L. — Scarlet pimpernel; pimpernel; poor man's weather-glass

---

**FAMILY:** Primulaceae — the Primrose Family

Members of this family have regular, perfect flowers with superior ovaries; **flowers:** 5-merous; **petals:** 5, fused; **stamens:** opposite and upon the petals; **ovary:** 1-celled; **style:** 1; **fruit:** a capsule; **leaves:** simple, without stipules.

**PHENOLOGY:** Pimpernel flowers from June throughout August.

**DISTRIBUTION:** *Anagallis arvensis* is a weed of diverse situations: roadsides, gardens, lawns, pastures, meadows, and waste places.

**PLANT CHARACTERISTICS:** Pimpernel is an herbaceous annual **plant; stems:** 4-angled; **leaves:** small, opposite, sessile, 1-2 cm, underside with minute "pits" and "spots"; **flowers:** scarlet to brick-red, solitary in the axils; **corolla:** deeply 5-parted, giving the appearance of separate petals, lobes twisted in bud; **staminal filaments:** hairy; **fruit:** capsule, upper half dropping away for seed dispersal (circumscissile); **flowers:** open only in fair weather, quickly closing at the approach of summer storms or during cloudy weather.

**POISONOUS PARTS:** All parts are to be considered poisonous.

**SYMPTOMS:** Leaves can cause contact dermatitis. Although ingestion of the plants may cause poisonings, well-documented cases of poisonings are rare. Sheep feeding tests produced death in 2 days at concentrations of 2% of the animal's weight; later in the growing season toxicity could not be demonstrated. Symptoms of toxicity included depression, anorexia, and diarrhea; **lesions** included kidney, heart, and rumen hemorrhaging, congestion of lungs, and a pale, crumbling liver. Loss of 6 calves was once reported.

**POISONOUS PRINCIPLES:** The toxin(s) remains unknown.

**CONFUSED TAXA:** There are no herbaceous annual plants with opposite leaves, scarlet flowers, and circumscissile capsules except *Anagallis arvensis.*

**SPECIES OF ANIMALS AFFECTED:** This plant is potentially poisonous to all species of animals.

**TREATMENT:** (11a)(b);(26)

**OF INTEREST:** Flowers are rarely white or sky-blue. Because flowers open and close in response to weather conditions, one of this plants names is "poor man's weather-glass."

# Apocynum

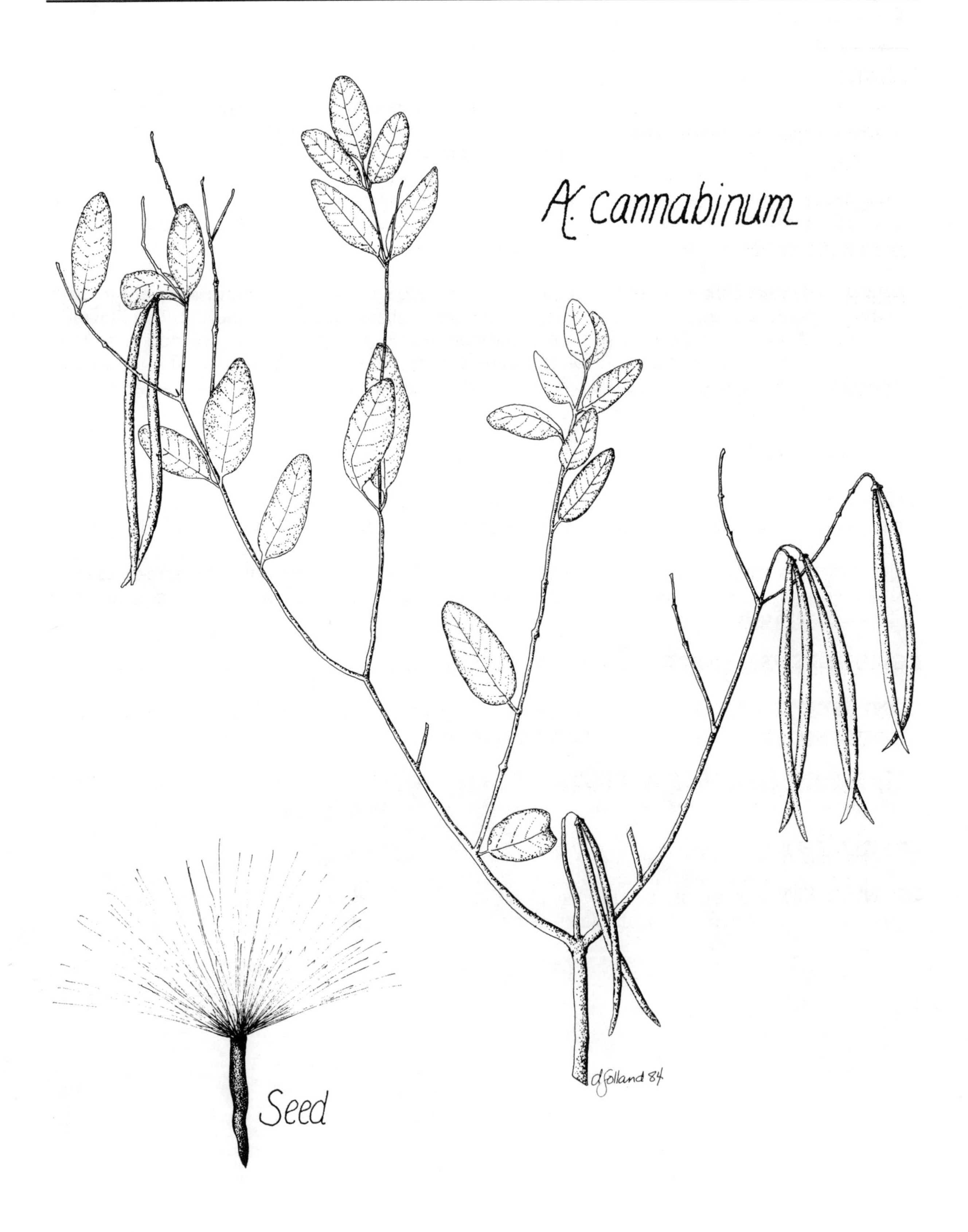

# GENUS: *Apocynum*

*Apocynum androsaemifolium* L. — Dogbane
*Apocynum cannabinum* L. — Indian hemp

**FAMILY:** Apocynaceae — the Dogbane Family

This family has opposite or alternate simple **leaves; flowers:** regular and perfect; **calyx:** deeply divided; **petals:** joined; **fruit:** 2 slender, many-seeded follicles; **sepals:** 5-lobed; **stamens:** as many as corolla lobes and alternate with them; **pistil:** 1.

**PHENOLOGY:** Flowering in June through September.

**DISTRIBUTION:** Found in open areas and in coarse soil and/or along streams.

**PLANT CHARACTERISTICS: Flowers:** erect; **calyx:** lobes usually taller than the middle of the corolla tube; **corolla:** petals are white to greenish white in *A. cannabinum* and pink in *A. androsaemifolium;* **fruit:** 10-15 cm long with 2-3 cm coma.

**POISONOUS PARTS:** Vegetative parts and the follicles.

**SYMPTOMS:** Little is known with respect to humans and livestock. One of the toxic glycosides, apocynamorin, when injected into a cat markedly raised the blood pressure. In another case, oral administration of some resinoid fractions to a dog produced gastric disturbance and death.

**POISONOUS PRINCIPLE:** Several cardioactive resins and glycosides.

**CONFUSED TAXA:** The two species of *Apocynum* can be readily distinguished. *Apocynum androsaemifolium* has pink corollas, 6-10 mm; *A. cannabinum* has white to greenish-white corollas, 3-6 mm.

**SPECIES OF ANIMALS AFFECTED:** Because the plant is distasteful to animals, incidences of poisoning are rare.

**TREATMENT:** (11a)(b); (26)

**OF INTEREST:** *Nerium oleander* (oleander), a poisonous evergreen shrub in this family is grown in more tropical climes and can be found in greenhouses in Pennsylvania. It contains the cardiac glycosides oleandroside, oleandrin, and nerioside. Symptoms include local irritation to the mouth and alimentary canal, vomiting, cramps, bloody diarrhea, dizziness, slowed pulse, irregular heartbeat, drowsiness, unconsciousness, respiratory collapse, and death.

# Arctium

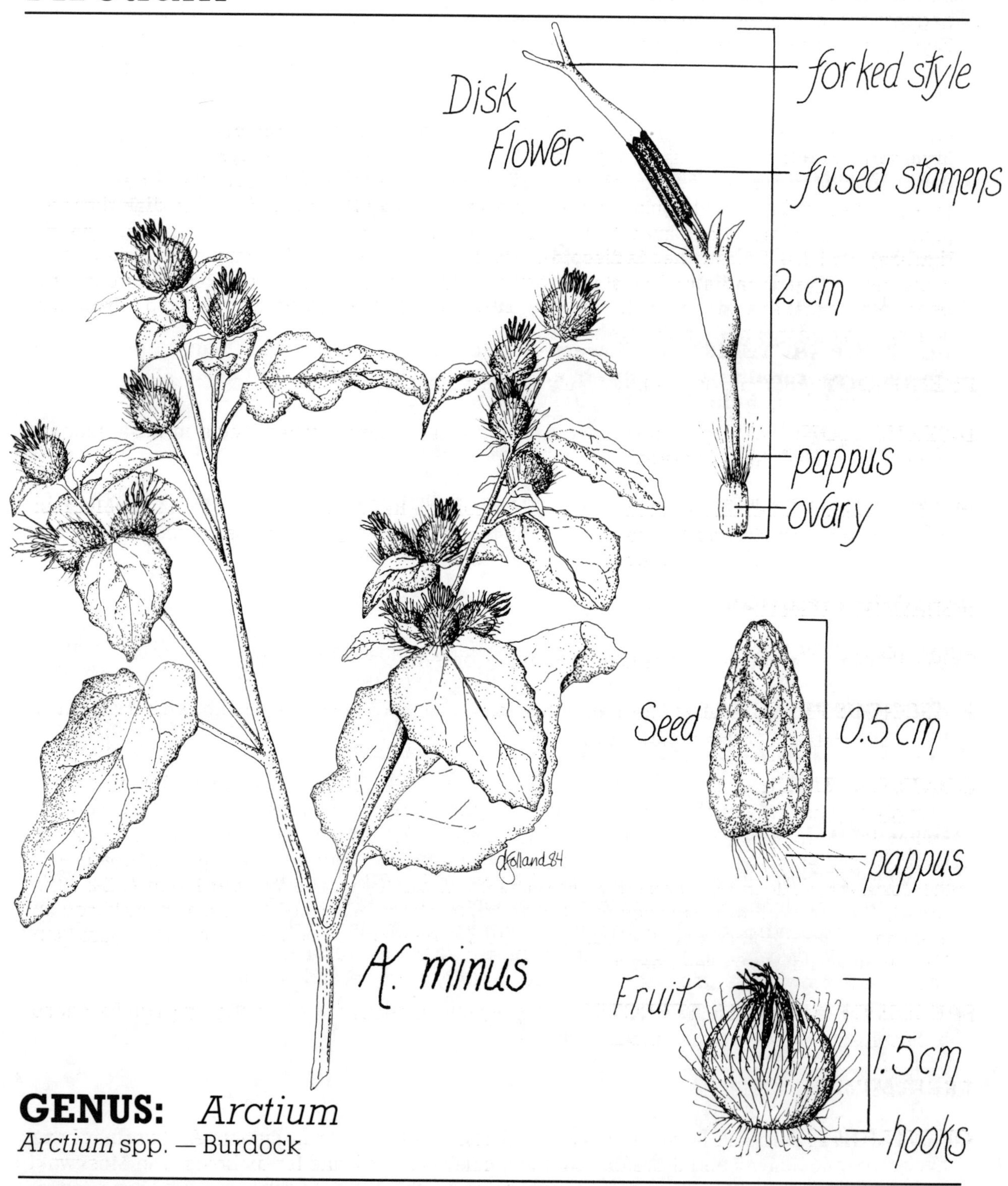


**GENUS:** *Arctium*
*Arctium* spp. — Burdock

**FAMILY:** Compositae (Asteraceae) — the Daisy Family

The largest family of vascular plants, the Compositae are distributed worldwide and are economically very important. Food members of the family include lettuce *(Lactuca);* endive, escarole, and chicory *(Cichorium);* artichoke *(Cynara);* salsify *(Tragopogon);* and sunflowers *(Helianthus).* Numerous species are used as ornamentals; especially notable are *Aster, Chrysanthemum,* daisys, cosmos, dahlia, strawflowers, cineraria, marigolds, zinnia, globe thistle, and edelweiss. Composite weeds often have detrimental impact. A nonexhaustive list would include ragweed, various thistles, horse weed *(Conyza),* galinsoga, fleabane, goldenrod, beggarticks, sowthistle, dandelion, and a host of less numerous taxa. A small proportion of the Compositae are poisonous and are detailed as separate entries (see *Arctium* spp., *Eupatorium rugosum, Helenium autumnale, Tanacetum vulgare,* and *Xanthium* spp.). *Senecio* is mentioned at the end of this entry.

Flowers of the Compositae are aggregated in close **heads,** on a **receptacle,** and surrounded by **involucral bracts** that are usually green. The **ovary** is inferior. The **calyx** is modified into a pappus, which crowns the summit of the ovary in the form of bristles, awns, scales, or teeth, or is absent. The corolla is either ligulate (flat and strap-shaped **ray-flowers),** or tubular (**disk-flowers,** often opening to form a 5-pointed star). The heads can be composed of all ray-flowers **(heads ligulate)** or all disk-flowers **(heads discoid),** or with ray-flowers along the margin and disk-flowers in the center **(heads radiate).** The **stamens** generally number 5, are fused upon the corolla, and bear anthers that are united into a tube. The **style** is 2- cleft; the ovary matures into a **fruit,** the achene, which contains a single, erect seed.

**PHENOLOGY:** *Arctium* species flower July through October.

**DISTRIBUTION:** The four *Arctium* species found in the Commonwealth occur in waste places, disturbed habitats, and roadsides.

**PLANT CHARACTERISTICS:** In the genus *Arctium* the **heads** are entirely discoid; **involucres:** globular; **involucral bracts:** attenuate to long, stiff, hooked tips; **pappus:** numerous, rough, separate, deciduous bristles; **leaves:** large and coarse.

**POISONOUS PARTS:** The green, above-ground portions may cause contact dermititis.

**SYMPTOMS:** Contact may cause itching, burning, or reddening of the skin.

**POISONOUS PRINCIPLES:** The agents that cause rash upon contact are unknown but probably are lactones (perhaps sesquiterpenes).

**CONFUSED TAXA:** Our four species of *Arctium* are: great burdock (*A. Lappa* L.), which has strongly angled leaf petioles with solid centers; hairy burdock (*A. tomentosum* Mill.), which has hollow leaf petioles and small flower heads (2.0-2.5 cm broad); the woodland burdock (*A. nemorosum* Lej. & Court.) has hollow leaf petioles and larger flower heads (2.5-3.5 cm broad); and the common burdock (*A. minus* (Hill) Bernh.), which has the smallest flower heads (1.2-2.5 cm broad). Several varieties of the common burdock, have been described by researchers based on leaf shape and corolla color. The burdocks are sometimes confused with the cockleburs (see *Xanthium*) and rhubarb (see *Rheum*).

**SPECIES OF ANIMALS AFFECTED:** Possibly only humans are susceptible to skin reactions from contact with the lactones in burdock.

**TREATMENT:** (23); (26)

**OF INTEREST:** Burdocks have been used in folk remedies for various ailments. *Arctium* may have hypoglycemic activity and therefore have potential as a medicine for diabetes. The Meskwaki Indians used *A. Lappa* root as an aid in childbirth, and 17th Century Europeans used it as a putative remedy for venereal disease.

Several species of *Senecio* (groundsel, ragwort), also in the family Compositae, produce pyrrolizidine alkaloids similar to those produced by *Crotalaria* (rattlebox), in the bean family. Of the *Senecio* species suspected or known to be toxic, only *S. vulgaris* L. occurs in Pennsylvania and its toxicity has not been proven in North America. The toxic groundsels are a problem mainly on western rangelands. *Senecio* toxicosis is similar to that described for *Crotalaria*.

# Arisaema

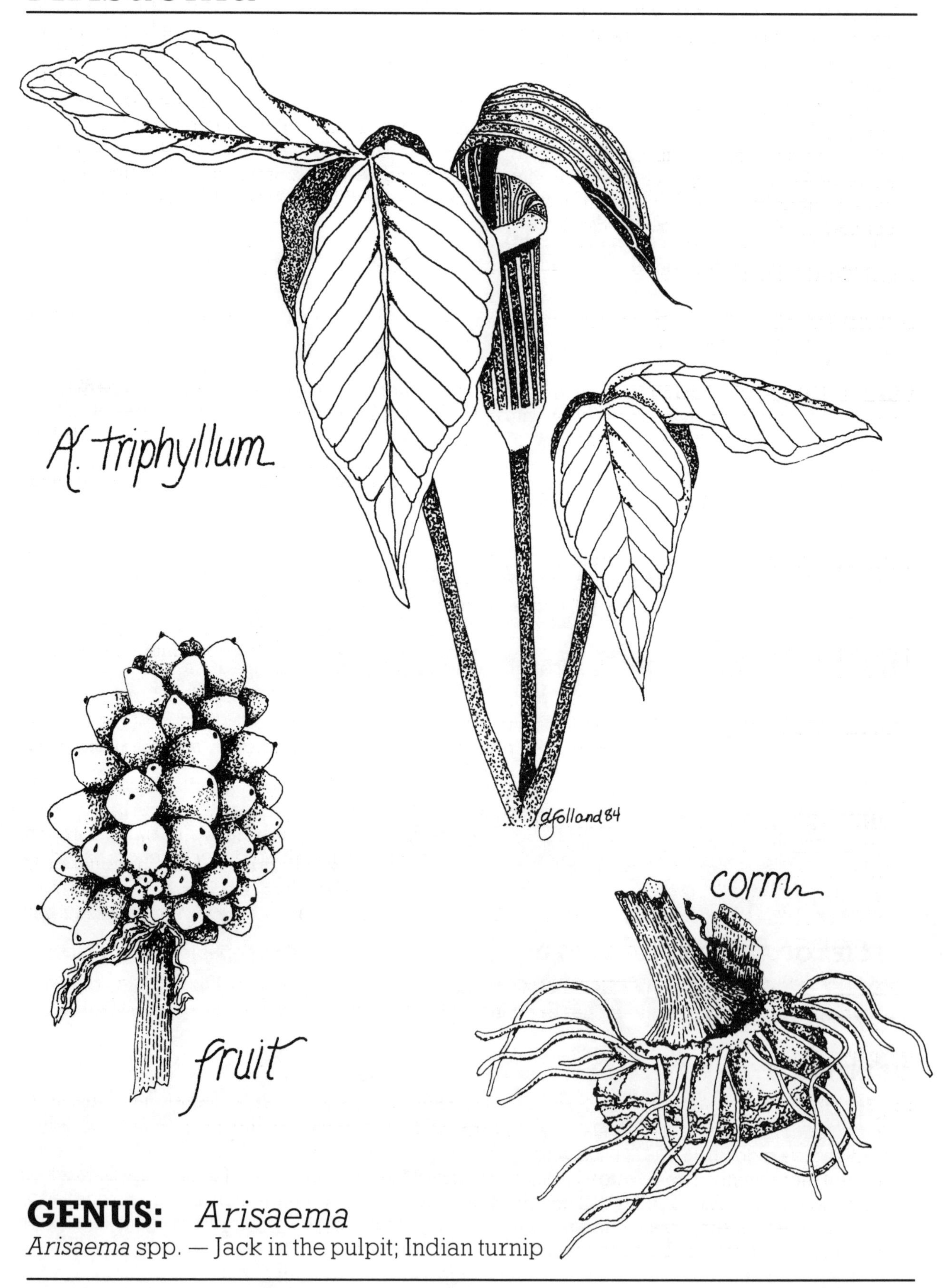


**GENUS:** *Arisaema*

*Arisaema* spp. — Jack in the pulpit; Indian turnip

**FAMILY:** Araceae — the Arum Family

The Araceae are primarily tropical in distribution. Seven genera, all but one containing a single species, are encountered in the flora of Pennsylvania. Members of the family are best known as house plants in our region (e.g. *Monstera, Philodendron, Anthurium, Dieffenbachia, Pothos, Scindapsus, Calla, Caladium,* and *Aglaonema).* The family is characterized by plants with milky, watery, or sharply pungent sap and calcium oxalate crystals in the tissue. The **flowers** are often unisexual. In some species, both male and female flowers occur in the same inflorescence. In other species, the plants bear either all male (staminate) or all female (pistillate) flowers. Regardless, the flowers are generally small and aggregated in a cluster on a thick, fleshy spike called a **spadix.** The spike is often surrounded or subtended by a bract or leaflike structure, the **spathe,** which may be colored and flowerlike.

**PHENOLOGY:** The species of *Arisaema* flower from late April to late June.

**DISTRIBUTION:** The plants are most commonly encountered in rich woods, thickets, moist areas, swamps, bogs, and swales.

**PLANT CHARACTERISTICS:** In *Arisaema* the flowers occur at the base of the **spadix;** the **spathe** is green or purple-brown, often with light-green lines. **Flowers:** small, in clusters, without a perianth; **staminate flowers:** composed of 2-5 subsessile anthers, opening at the apex; **pistillate flowers:** consisting of a 1-celled ovary and a broad stigmatic surface; **fruit:** a cluster of globose berries, red when mature, each containing 1-3 seeds; **leaves:** long-petioled, compound; **corms:** very acrid.

**POISONOUS PARTS:** Berries presumably are not poisonous but taste peppery. Leaves and roots (acrid sap) can cause contact dermatitis. Roots (corm) when eaten in quantity can cause severe burning in the throat and mouth. Inflammation can cause choking.

**SYMPTOMS:** Nausea, vomiting, diarrhea, and inflammation of the mucous membrane upon ingestion.

**POISONOUS PRINCIPLES:** The toxicosis is produced by mechanical piercing of the mucous membranes by calcium oxalate crytals, possibly a protein or asparagine, and other unknown toxins.

**CONFUSED TAXA:** There is some disagreement among botanists concerning the division of the genus *Arisaema* into species. Current research supports two species: *A. triphyllum* (L.) Schott. (with 3 varieties) and *A. Dracontium* (L.) Schott. The two are differentiated by the number of leaflets per leaf and the nature of the spadix. In *A. triphyllum* there are 3 leaflets and a blunt spadix covered by the spathe, whereas *A. Dracontium* leaves are composed of 7 to 13 leaflets and the spadix is long, protruding from the spathe.

**SPECIES OF ANIMALS AFFECTED:** Mortality in humans and livestock has not been reported in the literature; however, death has been induced experimentally in animal feeding studies.

**TREATMENT:** Dermatitis - (4); (23); ingestion - (11a)(b); (4); (6); (9); (10); (26).

**OF INTEREST:** Skunk cabbage, *Symplocarpus foetidus* (L.) Nutt., also a member of the Araceae, contains calcium oxalate crystals. An overdose of the underground parts causes nausea, vomiting, vertigo, disturbed vision, and headaches. Both *Arisaema* and *Symplocarpus* have been utilized medicinally. The Pawnee Indians pulverized the dried corms and dusted the powder on the head and temples to relieve aches. *Symplocarpus* "roots" have been dried and powdered to give the aged relief from asthma and catarrh. Some cultivated, poisonous plants of this family are discussed under the entries *Dieffenbachia* and *Philodendron.*

# Asclepias

# GENUS: *Asclepias*

*Asclepias* spp. — Milkweed

**FAMILY:** Asclepiadaceae — the Milkweed Family

The Asclepiadaceae is a family of succulent plants with milky sap. The milkweeds, which are known to most residents of the state, have a highly specialized flower, only briefly described here. The staminal filaments are basally fused into a tube, united to the corolla tube, and bear a whorl of appendages, collectively called the **corona.** The pollen of each anther is aggregated into a waxy mass, the **pollinium.** The **ovary** is superior, composed of 2 carpels, free at the base but fused at the apex into a common stigma; **fruit:** a pair of follicles, or 1 by abortion; **seeds:** flat, winged, with a tuft of hairs; **leaves:** opposite or whorled, simple, entire, without stipules; **inflorescence:** cymose, often appearing umbelliform; **flowers:** bisexual, 5-merous.

**PHENOLOGY:** The milkweeds generally flower June through August. A few species flower slightly earlier or extend slightly later into the season.

**DISTRIBUTION:** Some milkweeds are plants of wet places, swamps and bogs, others live in dry, rocky soil, while still others are weeds of cultivated fields, roadsides, pastures, and waste places.

**PLANT CHARACTERISTICS:** Those characteristics provided in the family description will identify the milkweeds. Generally the genus may be divided into those plants with narrow, lanceolate leaves and those with broad leaves and nearly parallel margins.

**POISONOUS PARTS:** The entire plant is considered poisonous.

**SYMPTOMS:** Toxicosis includes depression, weakness, staggering, tetanic seizures, elevated temperature, respiratory difficulties, dilated pupils, coma, and death. **Postmortem: gross lesions:** congestion of lungs, liver, and kidneys; acute catarrhal gastroenteritis; terminal dilitation of the heart ventricles; signs of central nervous system involvement, and atonic crop and gizzard in fowl; **histological lesions:** some cellular degeneration, especially of the kidney, may be apparent.

**POISONOUS PRINCIPLES:** The agents responsible for toxicity are not fully characterized, but toxic resinoids, cardioactive glycosides, and other components are suspected. The milky latex, upon contact, may elicit an allergic reaction in some sensitive individuals.

**CONFUSED TAXA:** Most people readily recognize milkweeds; the plants superficially resemble dogbane (see *Apocynum*).

**SPECIES OF ANIMALS AFFECTED:** Turkey, chickens, sheep, goats, cattle, and horses are susceptible to milkweed toxins. Humans are also poisoned by the plants.

**TREATMENT:** (11a)(b); (26)

**OF INTEREST:** Historically, several species of milkweed have been used for medicine, including *Asclepias tuberosa* L. and *A. syriaca* L. They contain asclepiadin, asclepion (a bitter principle), tannin, and volatile oil.

# Brassica

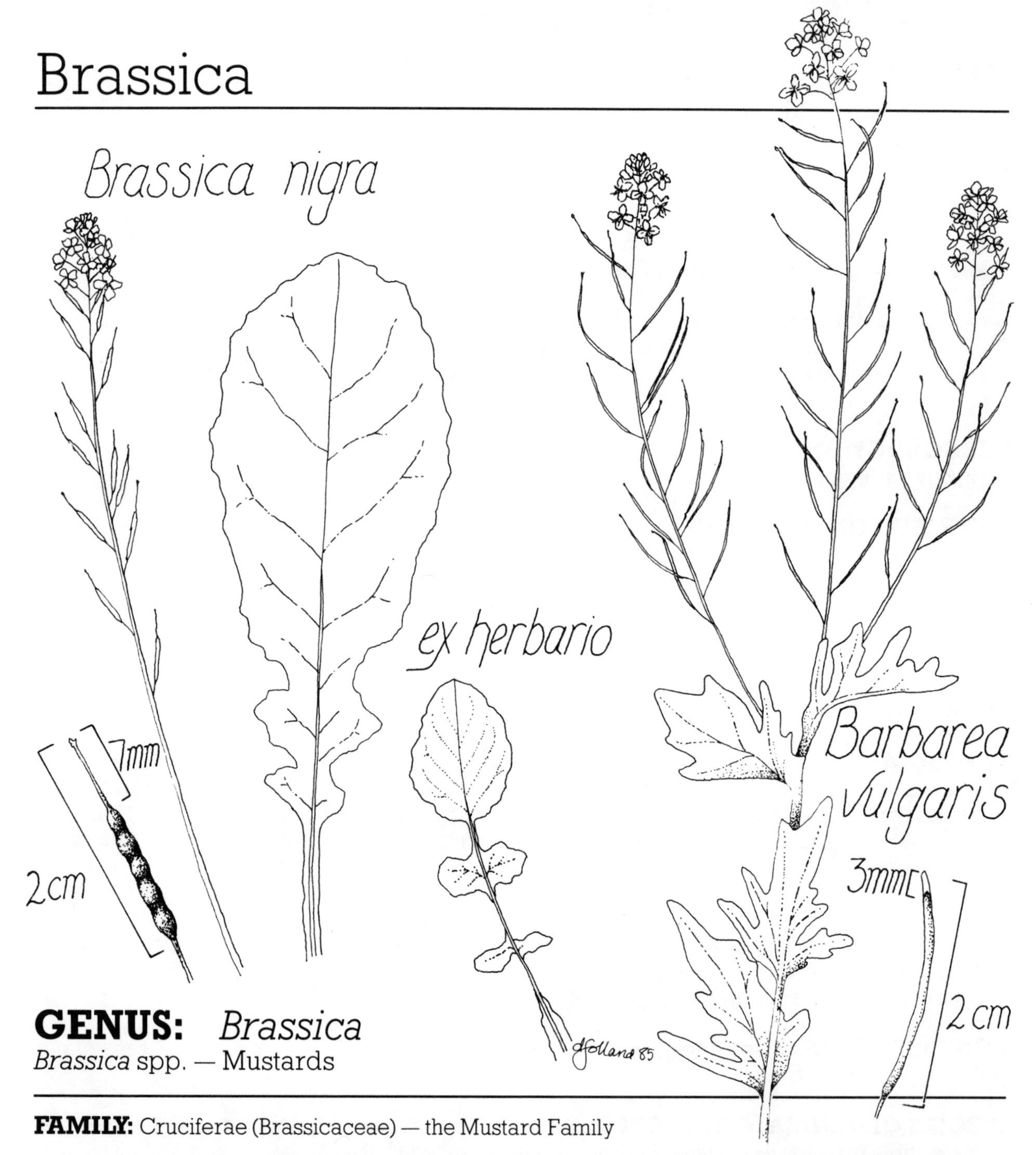


## GENUS: *Brassica*

*Brassica* spp. — Mustards

**FAMILY:** Cruciferae (Brassicaceae) — the Mustard Family

This is a large assemblage of pungent or acrid herbs of diverse growth habit. **Flowers:** regular and perfect, in terminal racemes or corymbs; **sepals:** 4, deciduous; **petals:** 4, limbs spreading to form a cross; **stamens:** 6, with 2 shorter and inserted lower than the other 4; **pistil:** 2 carpels; **ovary:** superior; **fruit:** a 2-celled capsule (a silique when elongated, a silicle when short and broad) usually opening by 2 valves from below; **seeds:** with a curved embryo important in taxonomic diagnosis; **leaves:** alternate, herbaceous without stipules.

This family includes many ornamental and important vegetable crops. A nonexhaustive list contains: *Brassica* (broccoli, Brussels sprouts, cabbage, Chinese cabbage, kale, kohlrabi, mustard, rutabaga, turnip), *Lepidium* (cress), *Nasturtium* (watercress), *Raphanus* (radish), *Armoracia* (horseradish), *Wasabia japonica* (Japanese horseradish), and *Crambe* (oil-seed).

**PHENOLOGY:** The genus *Brassica* contains several species that generally flower from May through October, depending on the taxon.

**DISTRIBUTION:** Some species of *Brassica* are cultivated plants; others are troublesome weeds of fields, waste places, gardens, and roadsides.

**PLANT CHARACTERISTICS:** Taxonomically the genus has been organized in several different fashions. Most members encountered in the Commonwealth bear saccate **sepals,** yellow **petals,** and have 4 rounded **staminal glands** at the base of the ovary. The **fruit** is terminated by a conspicuous **beak,** sometimes containing a basal seed.

**POISONOUS PARTS:** Seeds and plants with seed capsules are poisonous.

**SYMPTOMS:** The effects of *Brassica* poisoning vary depending upon the species of plant consumed. *Brassica Kaber* (DC) L. (charlock; wild mustard) is known to cause gastroenteritis, pain, salivation, diarrhea, and upper digestive tract disturbances, including irritation of the mouth. These symptoms also are associated with *B. hirta* Moench. (white mustard) ingestion. Some cultivated mustards such as *Brassica oleracea* var. *acephala* DC (common kale), *B. o.* var. *capitata* L. (cabbage), and *B. o.* var. *gemmifera* Zenker, (Brussels sprouts), cause hemolytic anemia and hemoglobinuria in some livestock. Goitrogenic substances (L-5-vinyl-2-thiooxazolidone) are known in kale, cabbage, and turnip (*Brassica rapa L.*).

**Postmortem: gross lesions:** (rape and kale) pulmonary emphysema; congestion and edema of lungs, alimentary tract inactivity causing gallbladder distension with viscid bile; **histological lesions:** rupture of pulmonary alveoli, emphysema and edema involving interlobular septa, tracheal and bronchial hemorrhages, mild toxic hepatitis, and centrilobular necrosis.

**POISONOUS PRINCIPLES:** The substance responsible for toxicosis is sinigrin, which in the presence of the enzyme myrosinase, is converted to glucose, allyl isothiocyanate (mustard oil), and potassium hydrogen sulfate. Mustard oils are poisonous. The toxicity, by ingestion, of allyl isothiocyanate has been determined (in cattle) to be 0.001% of the body weight. Also, mustards occasionally contain toxic concentrations of nitrate that may complicate toxicosis.

**CONFUSED TAXA:** There are 40 genera of mustards; many are yellow flowered. Botanical keys for the identification of mustards are complex and require mature fruits. One species frequently mistaken for a *Brassica* is *Barbarea vulgaris* R. Br. (Yellow rocket, winter cress), which also has been reported to produce mustard-oil type poisoning. One feature used to separate *Brassica* from *Barbarea* is the beak of the fruit: 8-15 mm long in *Brassica,* 1-3 mm long in *Barbarea*.

**SPECIES OF ANIMALS AFFECTED:** Reported poisonings include, cattle and sheep, *Brassica hirta* (white mustard); cattle and swine, *B. Kaber* (charlock); and ruminants, large quantities of *Brassica oleracea* var. *botrytis* (broccoli). Goiter formation is known for lambs (ewes) fed on *Brassica oleracea* var. *acephala* (kale) and rabbits fed *Brassica oleracea* var. *capitata* (cabbage).

**TREATMENT:** (26)

**OF INTEREST:** Numerous members of the mustard family have been reported to cause poisoning. Winter cress *(Barbarea vulgaris)* flowers April through June and is an abundant weed in Pennsylvania. One case was reported of a horse ingesting a relatively large amount of *B. vulgaris* and developing gastroenteritis. Rape (*Brassica campestris* L.), although a late fall pasturage crop, has been suspected of causing toxicosis. Horseradish (*Armoracia rusticana* P. Goertn.) has caused bloody vomiting and diarrhea in humans when consumed in large quantities. Loss of cattle, horses, and swine are known from the ingestion of vegetation and roots. Small children, who eat large quantities of raw mustard vegetables (cabbage, mustard, kale, Brussel sprouts, cauliflower, broccoli, rutabaga, turnip, radish, cress, horseradish and stock) can develop diarrhea and vomiting. Field penny-cress (*Thlaspi arvense* L.), a common weed of fields, roadsides, and waste places, is responsible for gastric distress in livestock. It has been suggested that toxicity in members of the mustard family increases after flowering. Additional plants suspected of being poisonous are *Erysimum* (wallflower), *Sisymbrium* (Tumbling mustard), *Descurainia* (Herb-Sophia), *Camelina* (False flax) and *Lepidium* (Peppergrass).

B. sempervirens
df

# GENUS: *Buxus*

*Buxus sempervirens* L. — Common boxwood, box

**FAMILY:** Buxaceae — the Box Family

This family contains two genera of plants used for ornamental purposes, *Buxus* and *Pachysandra*. The former is an evergreen shrub widely used in horticulture, while the latter is an evergreen ground cover, equally in common use. Characteristics for the family are; **flowers:** unisexual, regular, inconspicuous; **sepals:** 4, basally fused; **stamens:** 4, opposite the calyx lobes; **pistil:** 1; **ovary:** superior.

**PHENOLOGY:** Boxwood flowers in spring.

**DISTRIBUTION:** Cultivated as a hedge, foundation, specimen, or edging (dwarf) plant.

**PLANT CHARACTERISTICS: Leaves:** elliptic to lanceolate-oblong, broadest below the middle, dark green and lustrous above; **flowers:** in axillary clusters, with a terminal female flower and several male flowers below in the axils of bracteoles; **petals:** absent; **female flowers:** with a 3-celled **ovary; fruit:** a capsule with 3, two-horned valves.

**POISONOUS PARTS:** The leaves and stems are poisonous. Toxicity to horses is estimated to be 0.15% (green-weight basis) of body weight, which for an average animal is equivalent to 1.5 lbs of leaves.

**SYMPTOMS:** Severe gastroenteritis, vomiting, bloody diarrhea, stomach pains, convulsion, and death through respiratory failure may result from ingestion of boxwood.

**POISONOUS PRINCIPLES:** The akaloid buxene (buxine) has been implicated in poisonings. Other active principles are probably involved, including a volatile oil.

**CONFUSED TAXA:** No other cultivated plants have simple, opposite, oval, leathery leaves. Some varieties of holly *(Ilex)* or cotoneaster *(Cotoneaster)* may be confused with box, but these have alternate, not opposite leaves.

**SPECIES OF ANIMALS AFFECTED:** Horses, sheep, pigs, and cattle have been poisoned.

**TREATMENT:** (11a)(b); (26)

# Caltha

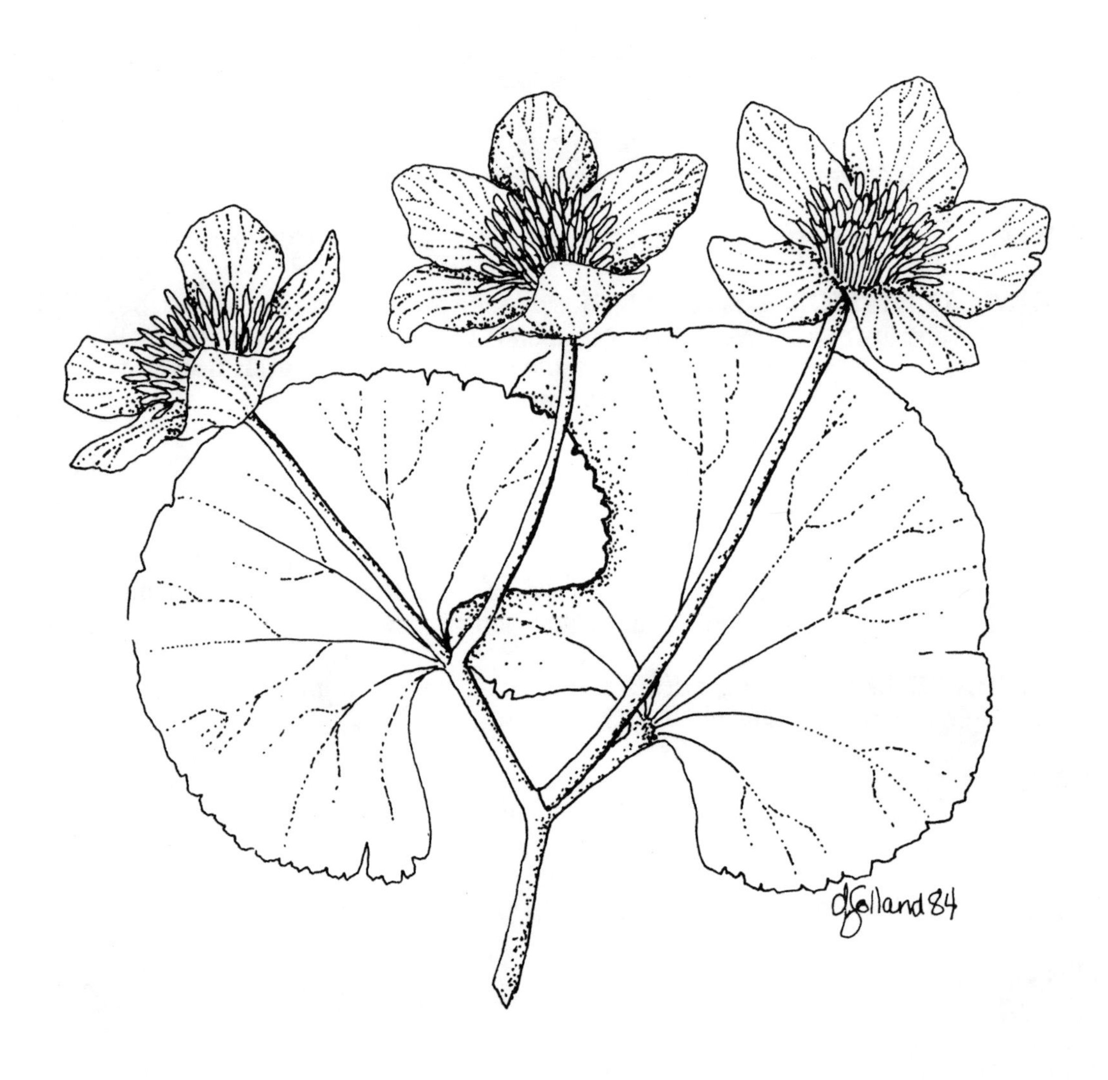

*C. palustris*

# GENUS: *Caltha*

*Caltha palustris* L. — Marsh marigold: cowslip

---

**FAMILY:** Ranunculaceae — the Buttercup Family (see *Actaea*)

**PHENOLOGY:** Marsh marigold usually flowers in April and May.

**DISTRIBUTION:** Marsh marigold, as the name implies, is often encountered in wet meadows, swamps, bogs, and shallow water. It can also occupy wet, shaded woodlands. Marsh marigolds will migrate onto low-lying areas of a lawn from an adjacent drainage ditch.

**PLANT CHARACTERISTICS:** Marsh marigolds have hollow **stems,** 2-6 dm tall, branched above; **basal leaves:** long-petioled, apically becoming progressively less petioled; **sepals:** 5-9, bright yellow and resembling petals. Each flower produces 4-12 **follicles,** 10-15 mm long.

**POISONOUS PARTS:** All parts of the mature plant are poisonous. Young plants are reported to be less toxic or not poisonous at all.

**SYMPTOMS:** Toxic principles can cause restlessness, depression, nervous excitation, stomach upset, salivation, weakness, and death.

**POISONOUS PRINCIPLES:** Toxins include the anemonin precursor proto-anemonin. The $LD_{50}$ i.p. in mice for anemonin is 150 mg/kg.

**CONFUSED TAXA:** *Caltha* resembles several of the poisonous buttercups in the genus *Ranunculus.* Buttercup flowers contain nectariferous spots or scales at the base of each petal, and the fruit is a single-seeded achene. *Caltha* is devoid of nectifers, and the fruit is a several-seeded follicle. There are several varieties of *Caltha palustris* differentiated by stem and leaf characteristics; all are considered poisonous.

**SPECIES OF ANIMALS AFFECTED:** Humans, cattle, and horses have been reported poisoned by ingestion of marsh marigold. Like buttercups, marsh marigold is acrid and not palatable. Because the poisonous agent is volatile, Marsh marigold in dried hay is reported harmless.

**TREATMENT:** (11a)(b); (26); (5)-2mg subcutaneously; (27); (6)

**OF INTEREST:** Many members of the family Ranunculaceae contain similar toxins (see *Ranunculus* and *Delphinium*).

# Campsis

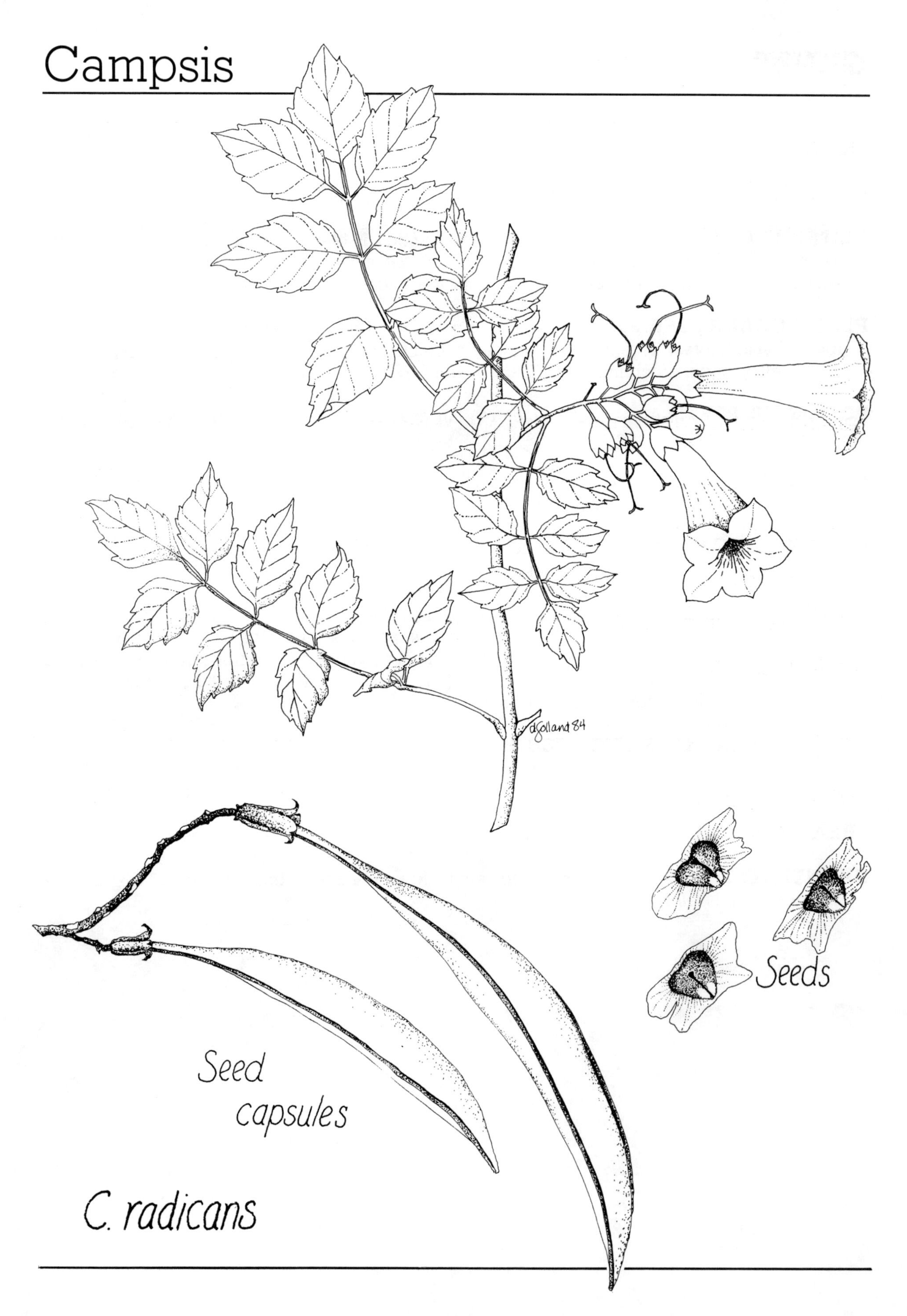

# GENUS: *Campsis*

*Campsis radicans* (L.) Seem. — Trumpet-creeper

---

**FAMILY:** Bignoniaceae — the Trumpet-creeper Family

The Bignoniaceae are a moderately large group of plants comprised of trees, shrubs, and woody vines. **Leaves:** ordinarily opposite; **flowers:** bisexual, irregular, 2-lipped, and often showy; **stamens:** borne on the petals, typically 4, in 2 pairs; **ovary:** superior, 2-celled; **fruit:** a 2-valved capsule; **seeds:** conspicuously winged.

**PHENOLOGY:** This plant blossoms during July and August.

**DISTRIBUTION:** Trumpet-creeper vines are showy plants that thrive in fertile soil in bright locations. They are occasionaly cultivated and frequently escape into moist woods, along fence-rows, and roadsides.

**PLANT CHARACTERISTICS:** A deciduous **shrub or vine** climbing by aerial roots; **leaves:** opposite, odd-pinnate; **leaflets:** toothed, 7-11; **flowers:** tubular, large, orange or scarlet, 6-8 cm long, produced in terminal, crowded inflorescences; **calyx:** unequally 5-toothed; **corolla:** 5-lobed, slightly 2-lipped.

**POISONOUS PARTS:** The leaves and flowers can cause contact dermatitis.

**SYMPTOMS:** Symptoms include skin inflammation, persistent blisters, and skin discomfort.

**POISONOUS PRINCIPLES:** Unknown

**CONFUSED TAXA:** No other woody vine in the Commonwealth possesses the plant characteristics described above.

**SPECIES OF ANIMALS AFFECTED:** People are reported to react on contact with trumpet-creeper. It is doubtful that animals are affected.

**TREATMENT:** (23); (4); (26)

# Caulophyllum

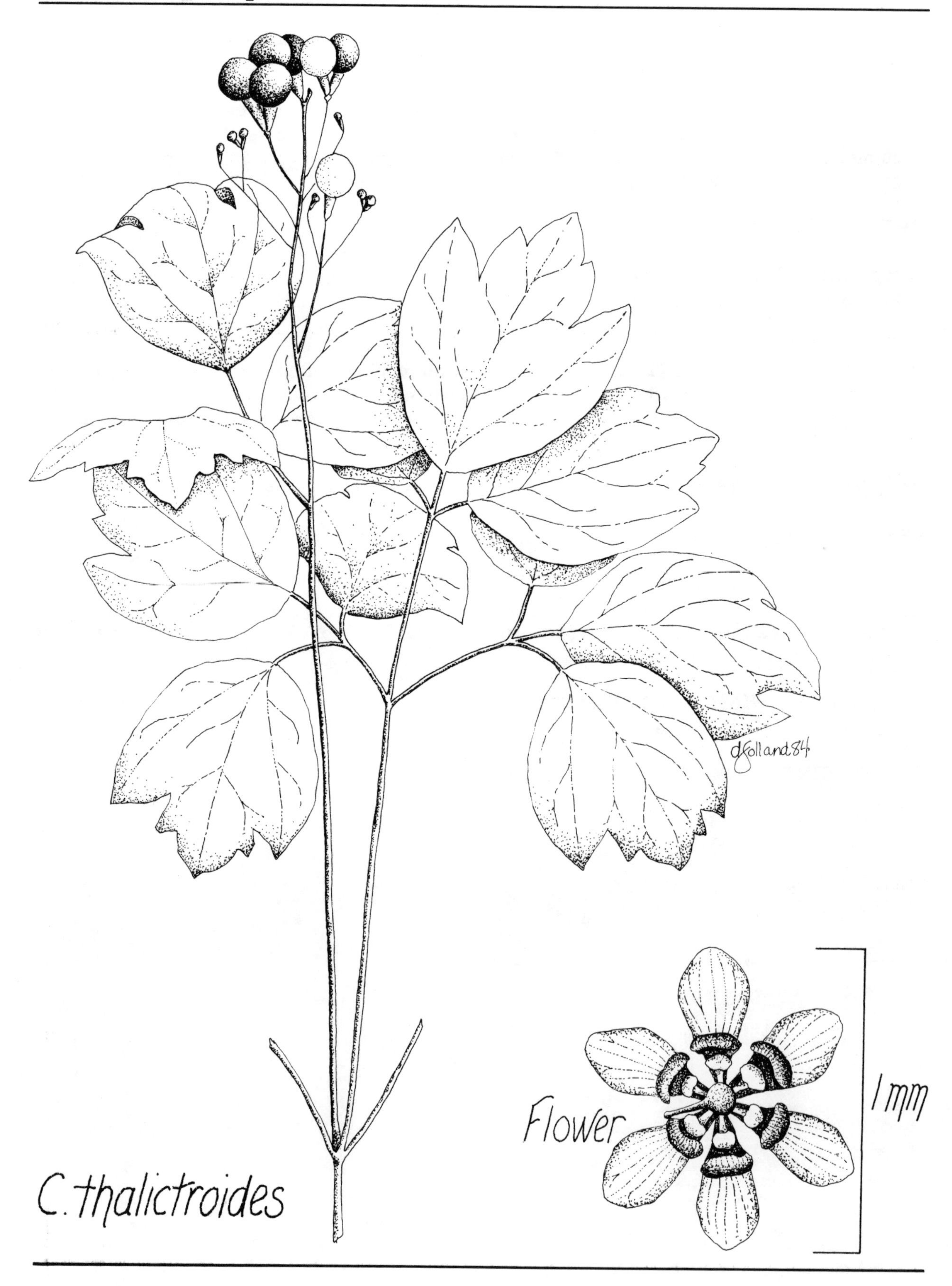

# GENUS: *Caulophyllum*

*Caulophyllum thalictroides* (L.) Michx. — Blue cohosh

---

**FAMILY:** Berberidaceae — the Barberry Family

Composed of about 12 genera and 200 species in the north temperate zone, this family is distinguished by two series of **stamens,** the outer whorl occurring opposite the petals. The calyx and corolla also may be in two whorls. The **anthers** shed pollen by valves, flowers possess a single **pistil,** and in most genera the stamen number is equal to the number of petals. The exception is *Podophyllum,* in which the stamen number is twice that of the petals. Economically the family contributes 13 genera of ornamental plants. Ripe fruits of *Podophyllum* (May apple) are edible (preserves and beverages) but leaves and roots are poisonous; *Berberis* (barberry) fruits are edible; *Caulophyllum* (blue cohosh) seeds are poisonous when raw but considered safe when roasted and used as a substitute for coffee beans. **Flowers** in the family have perfect symmetry, and free parts; **sepals:** 4 or 6; **petals:** as many as or more than the sepals, sometimes reduced to nectaries; **ovary:** 1, 1-celled, superior; **ovules:** 1-many.

**PHENOLOGY:** The single species flowers April and May.

**DISTRIBUTION:** Blue cohosh is common in the rich, moist deciduous woods of northeastern U.S.

**PLANT CHARACTERISTICS:** *Caulophyllum* is an erect, graceful perennial, from a small knotty rootstock, growing to 1 m high; glaucous when young; **flowers:** yellowish green or greenish purple with six petaloid **sepals,** subtended by 3-4 sepal-like **bracts;** the six **petals** are reduced to small glandlike bodies; **stamens:** 6; **seeds:** dark blue, on stalks as long as the seeds. The erect stem bears a single large, sessile, triternate **leaf** resembling 3 biternate leaves.

**POISONOUS PARTS:** The leaves and raw seeds contain toxins.

**SYMPTOMS:** Severe gastroenteritis and stomach pains result from consumption of the poisonous parts.

**POISONOUS PRINCIPLES:** The poisonous parts contain the alkaloid methylcytisine and various glycosides.

**CONFUSED TAXA:** The flower and seed characteristics described above make it unlikely that blue cohosh will be readily confused with other plants.

**SPECIES OF ANIMALS AFFECTED:** Although the plant is quite bitter, poisoning in children has been reported due to ingestion.

**TREATMENT:** (11a)(b); (26)

**OF INTEREST:** The North American Indians used a root extract of this plant as an abortifacient and to promote menstruation.

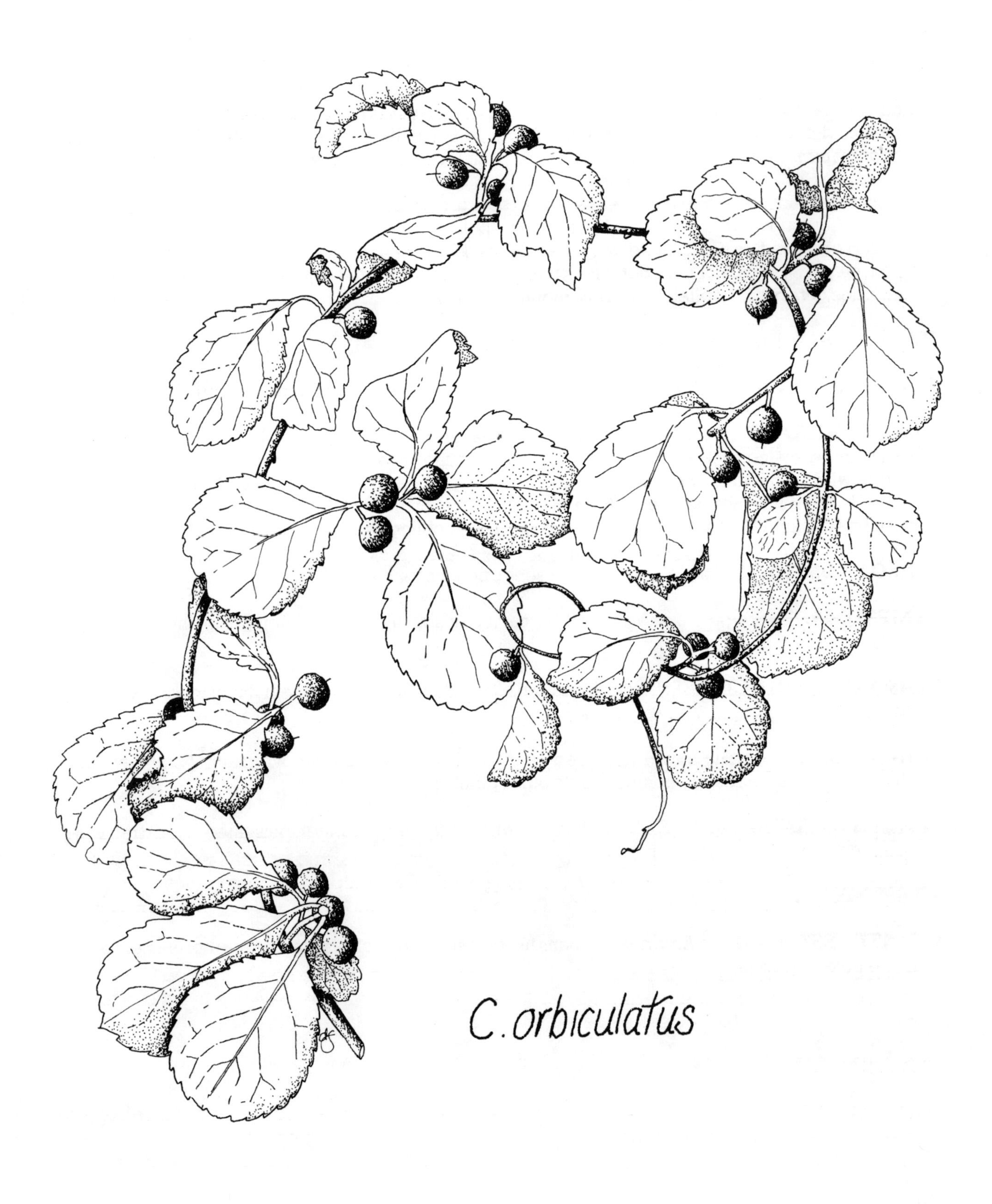
C. orbiculatus

# GENERA: *Celastrus, Euonymus*

*Celastrus* spp. — Bittersweet
*Euonymus* spp. — Wahoo; burning bush

**FAMILY:** Celastraceae — the Staff-tree Family

Represented in Pennsylvania by several native and introduced ornamental species, this family consists of trees, shrubs, or climbing vines. The **sepals, petals,** and **stamens** usually number 4 or 5. A **nectar disc** is present around the ovary. The **fruit** is often a capsule with the seeds wholly or partly surrounded by a fleshy, brightly colored membrane called the **aril.**

**PHENOLOGY:** Both *Celastrus* and *Euonymus* bloom May through June.

**DISTRIBUTION:** *Celastrus scandens* L. is a native plant of roadsides and woodlots, usually in rich soil. *Celastrus orbiculatus* Thunb., an introduced plant from eastern Asia, has escaped from sites where it is cultivated, especially in the southeastern corner of the Commonwealth and along the Susquehanna River drainage.

*Euonymus atropurpureus* Jacq. is native and found in moist woods; *E. europaeus* L. is a European native escaped from cultivation. *Euonymus* species are often cultivated as ornamental shrubs for their brilliant autumn foliage.

**PLANT CHARACTERISTICS:** *Celastrus* is **dioecious,** some plants producing only male flowers and others only female flowers. The whitish or greenish **flowers** are 5- merous; **fruits:** 3- valved, orange when mature, splitting to expose a fleshy red **aril.** The **vines** climb over vegetation and can be a tangling nuisance.

*Euonymus* plants are **shrubs** or **small trees** with **flowers:** perfect, 4- to 5-merous; **fruit:** 3- to 5-lobed, bright red or orange **aril.**

**POISONOUS PARTS:** Leaves, bark, and fruit are known to be poisonous in *E. atropurpureus* and *E. europaeus* and suspected to be poisonous in both *Celastrus* species.

**SYMPTOMS:** No cases of *Celastrus* poisoning could be located. Symptoms in *Euonymus* poisoning include: diarrhea, vomiting, unconsciousness, and mental disorders.

**POISONOUS PRINCIPLES:** Peptide and sesquiterpene alkaloids may be responsible for *Euonymus* toxicosis. The toxins act as violent purgatives.

**CONFUSED TAXA:** The seeds covered by brightly covered arils make these two genera distinct from other woody plants in the Commonwealth. *Euonymus* is a shrub or small tree with simple, opposite leaves, whereas *Celastrus* is a woody vine with simple, alternate leaves. The two *Celastrus* species can be distinguished by terminal panicles and leaves twice as wide as long in *C. scandens* and axillary cymes and broad leaves in *C. orbiculatus.*

**SPECIES OF ANIMALS AFFECTED:** Horses are recorded to have been poisoned by leaves, children by the fruits of *Euonymus.* Bittersweet is often brought into the home in autumn for its decorative fruits. Children should be alerted to the potential danger of this plant.

**TREATMENT:** (11a)(b); (26)

**OF INTEREST:** *Celastrus scandens* has been used in eastern North America as a folk medicine to treat vaginal discharge (leukorrhea). In other parts of the world, species in this genus have been used variously as abortifacients, stimulants, emetics, and cathartics, and the leaves chewed to relieve toothache.

# Chelidonium

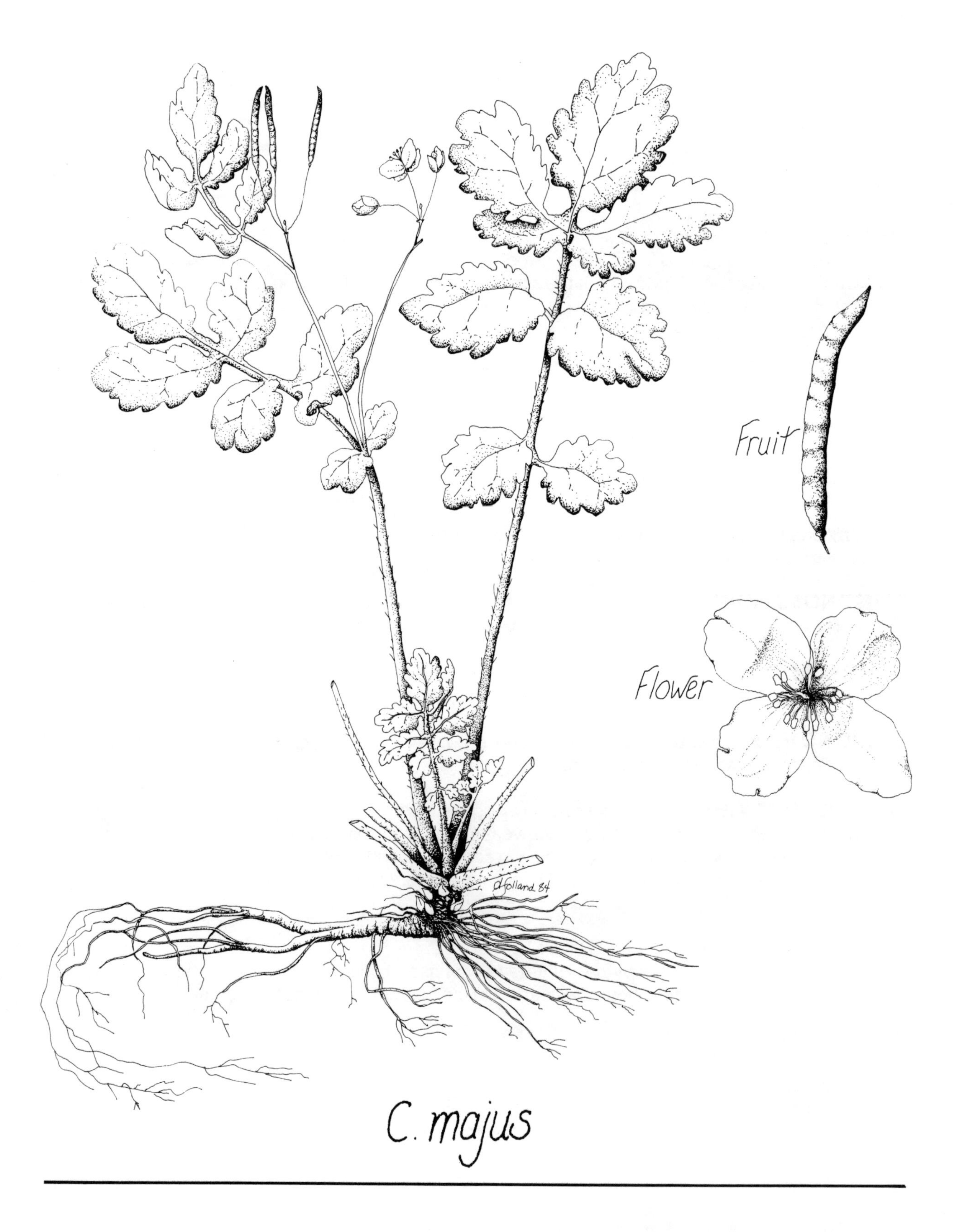

# GENUS: *Chelidonium*

*Chelidonium majus* L. — Celandine

---

**FAMILY:** Papaveraceae — the Poppy Family

**Flowers:** perfect, showy, bisexual and regular; **petals:** 4-8 or 8-12, separate, conspicuous, and deciduous; **sepals:** 2-3, falling early; **stamens:** very numerous; **ovary:** superior, 1-celled; **fruit:** a capsule, usually opening by valves or pores; **leaves:** exstipulate, alternate on the stems. When the integrity of the stem or leaves is disturbed, the wound produces a white, or colored milky **sap.**

**PHENOLOGY:** Celandine flowers in April through September.

**DISTRIBUTION:** Found in moist soil, gardens, rich woods, and roadsides were vegetation is dense.

**PLANT CHARACTERISTICS:** *Chelidonium majus* is a biennial or short lived perennial weed naturalized from Europe. **Sepals:** 2, falling early; **petals:** 4, yellow; **stamens:** numerous, with long, slender filaments and short, round anthers; **ovary:** glabrous, with a very short style with 2-lobed stigma; **flowers** small, several in a peduncled umbel; **leaves:** deeply lobed almost or quite to the midvein.

**POISONOUS PARTS:** The plant's sap, found in stems, roots, and leaves, is poisonous.

**SYMPTOMS:** If ingested, celandine produces severe gastroenteritis. Also, there is possibly some risk of skin irritation on contact.

**POISONOUS PRNCIPLE:** The toxins are alkaloid compounds, i.e. chelidonine, chelerythrine, protopine, sanguinarine, berberine, tetrahydrocoptisine, and others.

**CONFUSED TAXA:** This is the only naturalized weed having flowers with 4, brilliant yellow-orange petals and milky sap when bruised.

**SPECIES OF ANIMALS AFFECTED:** Apparently all species of livestock and humans develop gastroenteritis upon ingestion of celandine.

**TREATMENT:** (11a)(b); (26)

# Cicuta

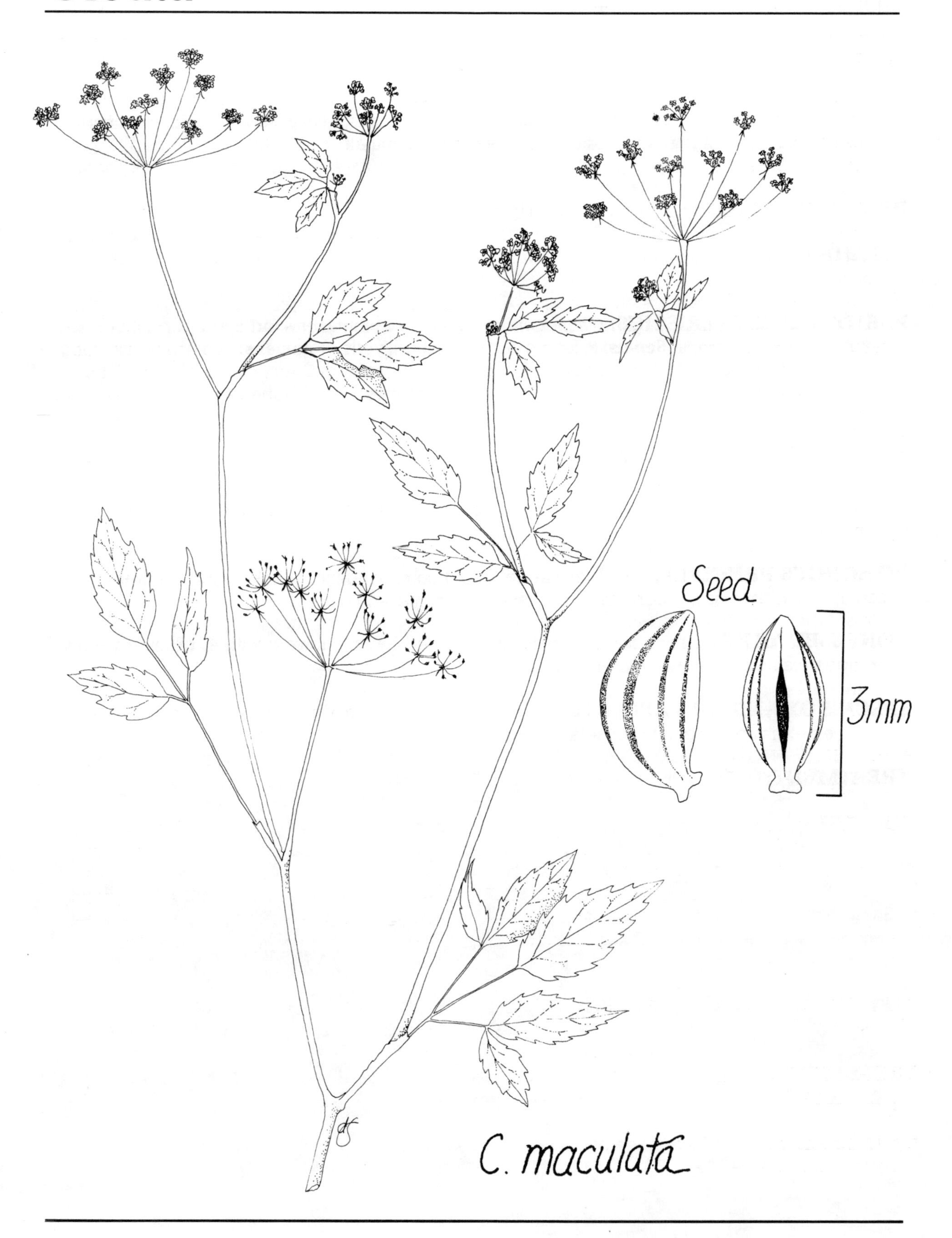

# GENUS: *Cicuta*

*Cicuta maculata* L. — Water hemlock; cowbane; beaver poison

---

**FAMILY:** Umbelliferae (Apiaceae) — The Umbel Family

This group consists of herbs with **flowers:** usually regular and perfect, in a simple or compound (most common) umbel; **stamens:** 5, inserted on a disk; **styles:** 2; **ovary:** 1, inferior; **leaves:** alternate or basal, the petiole sometimes bearing a basal sheath.

**PHENOLOGY:** Water hemlock flowers June through August.

**DISTRIBUTION:** Found in marshes, swamps, ditches, streams, and marshy meadows and pastures.

**PLANT CHARACTERISTICS:** **Flowers** are in compound umbels; **sepals:** triangular; **petals:** white; **stem:** glabrous, jointed with hollow internodes, base of stem swollen and transversely partitioned; **root:** tuberous; **leaves:** pinnately compound with well-defined leaflets; **leaflets:** linear to lance-ovate, 3-10 cm long; primary lateral veins in the leaflets are directed to the crotch of the teeth; **umbels:** numerous, 5-12 cm wide.

**POISONOUS PARTS:** All parts are extremely poisonous. A piece of root the size of a pea is sufficient to kill a human. A piece of root the size of a walnut will kill a cow in fifteen minutes, and about 1 lb of dried plant may kill a horse.

**SYMPTOMS:** Usually within ½ hour after ingesting a lethal dose the following symptoms occur: excessive salivation, then tremors and spasmodic convulsions with intermittent relaxation (the convulsions are extremely violent). Abdominal pain is evident, pupils are dilated, and temperature may be several degrees higher than normal. Humans may become delirious. Nausea and vomiting occur if the animal can vomit. Bloating is common. Additional symptoms include diarrhea, irregular pulse and heart rate, and behavioral abnormalities such as rolling of the eyes, turning in circles, twisting of the neck, falling down, and opening and shutting of the mouth. Death is due to respiratory failure after complete paralysis. **Postmortem: gross and histological lesions:** no obvious changes.

**POISONOUS PRINCIPLE:** Cicutoxin, a highly unsaturated alcohol, is responsible for poisoning. It is usually associated with the yellowish, oily liquid located in the lower stem and roots.

**CONFUSED TAXA:** Young plants of elderberry, *Sambucus* spp. (Caprifoliaceae), resemble water hemlock. The leaves are opposite in elderberry and alternate in water hemlock. Elderberry may be mildly toxic. A cyanogenic glycoside, as well as an alkaloid, are present in elderberry leaves, flowers, berries, and particularly the roots. In moderate amounts these substances are purgative. Fresh berries are paradoxical--harmless when cooked but sometimes producing nausea when uncooked. **Postmortem** evaluation of elderberry toxicosis reveals bright red blood characteristic of cyanide poisoning.

**SPECIES OF ANIMALS AFFECTED:** All species of animals and humans are affected by cicutoxin.

**TREATMENT:** (11a)(b); (26); Convulsions can be controlled by parenteral, short-acting barbiturates.

**OF INTEREST:** Cattle have been poisoned by drinking water from an area where water hemlock roots were trampled.

# Claviceps

## GENUS: *Claviceps*

*Claviceps* spp. — Ergot

**FAMILY:** Ascomycetes — the Ascomycete Family

"Sac fungi" produce spores in asci, or sacs. Some economically important fungi in this family include yeast and the edible morels. Each **ascus** produces a definite number of **spores,** usually eight, and the fungal threads have cross walls **(septae)** with a central perforation. The fungi responsible for Dutch elm disease, Chestnut blight, and a variety of human lung disorders are all ascomycetes.

**OCCURRENCE:** *Claviceps* parasitizes the ovary of grasses, especially rye, wheat (durum is most susceptible), barley, and some wild species. Infection occurs when host flowers begin to open.

**DISTRIBUTION:** Ergot occurs on pasture land grasses or hay and cereal grains from cultivated fields.

**POISONOUS PARTS:** The poisonous part is the **sclerotium** (ergot body), a grain-shaped mass that replaces the grass ovary. This varies in size from the same as the grain to 4 times larger. The fungal mass, homogeneous and white when cut open, is shed with the grass and acts as the overwintering phase of the fungus. Federal law prohibits use of cereal grains containing more than 0.3% sclerotia by weight.

**SYMPTOMS:** Two syndromes are produced by ingestion: 1) gangrenous, and 2) convulsive. Ingestion of small amounts daily over a short period results in necrosis of tissues in the extremities, producing dry gangrene. Gangrene is caused by constriction of the blood vessels with blockage of circulation. This results in lameness, coldness, and insensitivity to pain of the affected part. In some instances serum seepage can cause secondary infection, which may be associated with nausea, vomiting, abdominal pain, and constipation or diarrhea. Pregnant animals spontaneously abort. Mucus membranes of the oral cavity may be inflamed or damaged. In humans gastrointestinal distress and headache may be present. Fowl may lose their combs and beaks. Convulsive ergotism results from ingestion of large quantities of ergot. In addition to the above syndrome, nervous symptoms appear, which are characterized by hyperexcitability, paranoia, rapid pulse, and belligerence. In livestock, death may result from dehydration or starvation within a few days or a month. In humans, whole body spasms and delirium may be present.

**Postmortem: gross lesions:** dry gangrene of ears, limbs, and tail; moist gangrene of feet and phalanges; inflammatory zone between gangrenous and living tissue; visceral organ congestion and hemorrhage may be present; **histological lesions:** dry gangrene shows coagulated blood, bacterial infection; moist gangrene shows coagulation and liquefaction necrosis in which large bacilli are evident.

**POISONOUS PRINCIPLES:** Alkaloids, amines, and other organic compounds are present in ergot. The antihemorrhagic alkaloids probably are the major problem. Chemical formulas are known for two dozen alkaloids, derivatives of lysergic acid. Compounds include ergocryptine, ergocornine, ergocristine, ergotamine, ergosine, and ergonovine.

**CONFUSED TAXA:** Identification of parasitic fungi should be made by trained specialists.

**SPECIES OF ANIMALS AFFECTED:** Ergotism in animals other than cattle and humans is rare.

**TREATMENT:** (11a)(b); (26)

**OF INTEREST:** The scientific and herbal literature contains much material on ergot, including the natural history and biology of the fungus, chemistry and physiology of its alkaloids, disease symptoms, and use of ergot in medicine. The mode of action of ergot is to stimulate smooth muscle. Because of the persistent contraction of smooth muscle blood vessels, dry gangrene of the extremities occurs.

The infection of grain seed heads by fungus is not uncommon. Cultivated barley, *Hordeum vulgare* L., is subject to infection by *Gibberella saubinetti* (imperfect stage is *Fusarium graminearum)* and becomes toxic to certain species of animals. The severity ranges from vomition and listlessness in pigs to no apparent effects in ruminants.

*Festuca arundinacea* Schreb. (=*F. elatoir* var. *arundinacea* (Schreb.) Wimmer) may cause ergotlike poisoning in livestock, especially cattle (horses appear to be immune). The toxicosis may be due to parasitic fungus rather than the grass itself. The cattle disease "fescue foot" resembles gangrenous ergotism. **Postmortem: gross lesions:** edema followed by necrosis, distal to a line of demarcation, in extremities; extremities dry, shrivel, and separate; **histological lesions:** arterioles at the coronary band have thickened walls and constricted lumens.

# Codiaeum

*C. variegatum*

# GENUS: *Codiaeum*

*Codiaeum variegatum* (L.) Blume — Variegated laurel; croton

**FAMILY:** Euphorbiaceae — the Spurge Family.

This family of worldwide distribution contains many genera. Economically the family is very important for ornamental plants and for providing rubber, edible roots and fruits, and medicinals. Poisonous properties are also common in members of this family.

Plants in the Euphorbiaceae are diverse in appearance. Generally, **flowers:** regular, hypogynous, and unisexual; **calyx:** present or absent; **petals:** usually absent; **stamens:** 1 to many, sometimes with branched filaments; **ovary:** superior, usually 3-celled; **fruit:** a capsule splitting into 3, 1-seeded sections. Plants are often succulent and cactuslike with milky or watery **juice.**

The type genus, *Euphorbia,* has a compact inflorescence called a **cyathium,** a structure simulating a complete flower. The individual "flower" is actually a floral-cluster, usually consisting of 1 female with a single pistil and several male flowers each with one stamen, all on a jointed pedicel. The flowers of the cyathium are enclosed in a cuplike **involucre,** which often contains glands and appendages and may be subtended by brightly colored **bracts** (e.g. the red "petals" of a poinsettia "flower").

**PHENOLOGY:** *Codiaeum* are house plants used in interior decorating. Under optimal conditions, they may flower indoors in our northern climate.

**DISTRIBUTION:** Crotons are grown under glass around office and home for their colored ornamental foliage. These plants are native to the Malay Peninsula and Pacific Islands.

**PLANT CHARACTERISTICS:** *Codiaeum* has the following characteristics; **leaves:** alternate, simple, rarely lobed, leathery, glabrous, petioled, and variously colored; and **flowers:** small, in axillary racemes.

**POISONOUS PARTS:** Leaves, stems, and flowers are considered potentially poisonous. Some species in the genus are edible; young, yellow varieties eaten in the East Indies are reported to be sweet. House plant varieties contain irritant juices that may be noxious or allergenic. Croton juices are used medicinally as purgatives, abortifacients, sudatories, and antitussives.

**SYMPTOMS:** Digestive upset results from ingestion. Allergic reactions may occur upon contact.

**POISONOUS PRINCIPLES:** The toxins are currently unidentified. Some plants may contain caustic latex. The plant sap also contains 6-8% tannin.

**CONFUSED TAXA:** The common name croton (genus *Codiaeum*) should not be confused with the genus *Croton.* Plants in the former genus are glabrous, whereas in the latter they bear stellate pubescence.

**SPECIES OF ANIMALS AFFECTED:** Because the variegated laurels are house plants, probably only children and pets are potentially at risk.

**TREATMENT:** (11a)(b); (26); (1)

**OF INTEREST:** Many members of this family are moderately to severely poisonous. *Croton,* mentioned above, is occasionally encountered in dry, sandy soil and waste places in Pennsylvania. It is known to contain acrid, irritant principles.

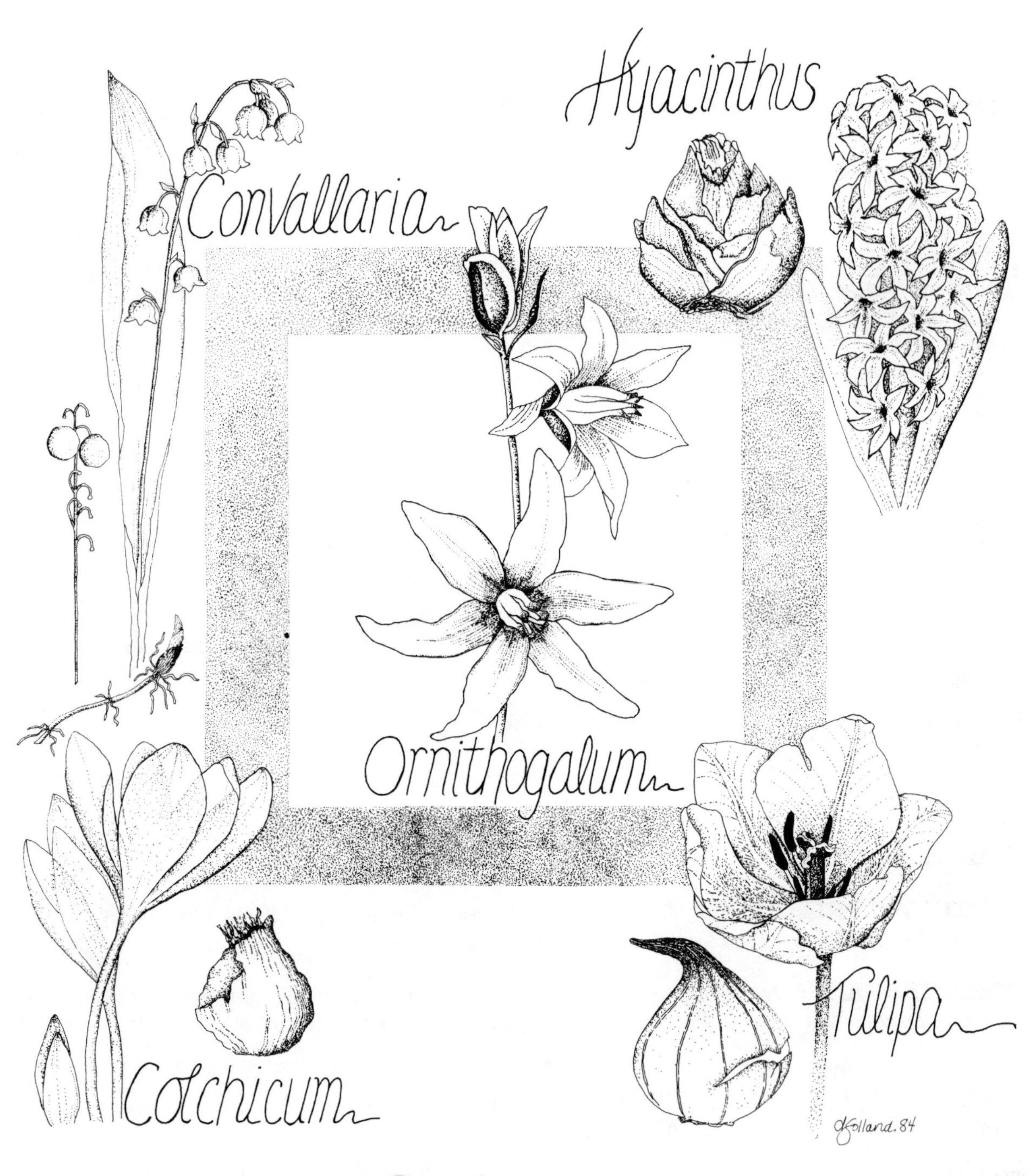
Hyacinthus
Convallaria
Ornithogalum
Tulipa
Colchicum
Holland.84

Liliaceae

# GENUS: *Colchicum*

*Colchicum autumnale* L. — Autumn crocus

---

**FAMILY:** Liliaceae — the Lily Family (see *Amianthium*)

**PHENOLOGY:** The white to light-rose or light-purple flowers appear in the fall.

**DISTRIBUTION:** Autumn crocus is cultivated around homes and in gardens. It rarely escapes and becomes naturalized.

**PLANT CHARACTERISTICS:** The **flowers** of *Colchicum* are chalice-shaped, with **stamens:** 6; and **styles:** 3, long and slender. The large **leaves** appear in the spring with the previous season's seed-pod and die back during summer.

**POISONOUS PARTS:** All parts are toxic, especially the bulb and seeds. Leaves are toxic at about 0.1% of an animal's weight.

**SYMPTOMS:** Toxicosis includes vomiting; purging; weak, quick pulse; gastrointestinal irritation; burning pain in mouth, throat, and stomach; and kidney and respiratory failure.

**POISONOUS PRINCIPLES:** The alkaloid colchicine and related compounds are responsible for poisonings.

**CONFUSED TAXA:** Spring crocus *(Crocus)* of the family Iridaceae can be confused with Autumn crocus. *Crocus* has 3 stamens and 1 style with 3 stigmas, whereas *Colchicum* has 6 stamens and 3 styles.

**SPECIES OF ANIMALS AFFECTED:** Children have been poisoned by eating the flowers; poisoning has been reported in all classes of livestock.

**TREATMENT:** (11a)(b); (22); (26)

**OF INTEREST:** The alkaloid is heat-stable and therefore not inactivated by high temperatures such as the ensilaging process. Livestock have been lost on ingestion of hay containing *Colchicum*. Milk from a lactating animal can poison the nursing offspring. The alkaloid is used medicinally as a gout suppressant, in the treatment of Familial Mediterranean Fever, in veterinary science as an antineoplastic, and in genetic research.

Other cultivated members of the Liliaceae are known or suspected to be poisonous. Tulip (*Tulipa* spp.) bulbs cause severe purgation in cattle. Hyacinth (*Hyacinthus* spp.) bulbs, if eaten in quantity, produce gastrointestinal upset, severe purgation, diarrhea, vomiting, and cramps. During World War II, bulbs were fed to cattle in the Netherlands, producing violent gastric reactions.

Still other Liliaceae of concern are: dogtooth violet (*Erythronium* spp.) - bulbs known to poison poultry; bunchflower (*Melanthium* spp.) - may poison sheep, cattle, and horses; fritillaria *(Fritillaria meleagris)* - contains a heart-depressant alkaloid; and squills *(Urginea maritima)* - bulbs contain cardiotonic glycosides having digitalislike action.

Additionally, some edible members of the Liliaceae have been reported to produce toxicosis. Cultivated onions (*Allium cepa* L.), chives (*Allium schoenoprasm* L.), and wild onion (*Allium canadense* L.) produce compounds known to be toxic in large quantities; wild garlic (*Allium vineale* L.) also is suspect. Symptoms may include anemia and intense gastroenteritis. When death follows, the animal tissue and even the necropsy room are permeated with onion odor. The compounds allicin and alliin, known to have antimicrobial properties, may be involved. Cultivated asparagus (*Asparagus officinalis* L.) is reported to have killed dairy cattle upon ingestion of mature plants. The red berries of asparagus are eaten both raw and cooked, but because some individuals are sensitive to the berries, this practice is to be discouraged.

# Conium

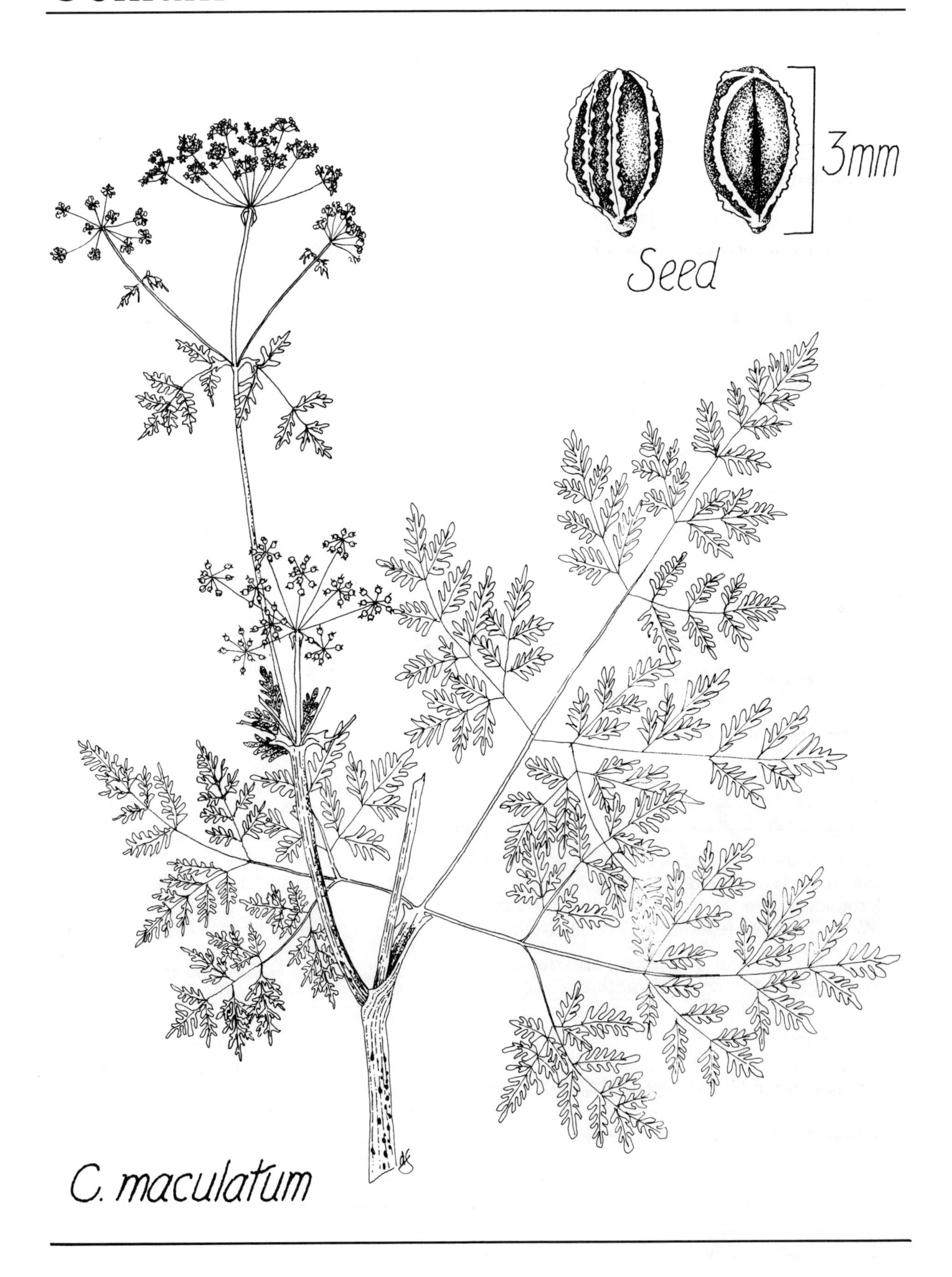

# GENUS: *Conium*

*Conium maculatum* L. — Poison hemlock; spotted hemlock; deadly hemlock; poison parsley

---

**FAMILY:** Umbelliferae (Apiaceae) — the Umbel Family (see *Cicuta*)

**PHENOLOGY:** Poison hemlock flowers June through September.

**DISTRIBUTION:** It is found in disturbed or waste areas such as roadsides and the edges of cultivated fields.

**PLANT CHARACTERISTICS:** Diagnostic features include, **stem:** purple-spotted, glabrous, and branched, up to 3 m tall; **leaves:** pinnately decompound, 2-4 dm long and toothed; flowering **umbel:** 4-6 cm wide (umbels are numerous); **fruit:** broadly ovoid, about 3 mm, laterally constricted; **petals:** white.

**POISONOUS PARTS:** All parts of *Conium maculatum* are extremely poisonous. Some studies reveal toxicosis at 0.25% (green-weight basis) of a horse's weight; 0.5% for a cow's. In contrast, experimental feeding studies on a cow showed symptoms at 2% of the animal's weight and produced death at about 4%.

**SYMPTOMS:** The symptoms, in order of appearance are: nervousness, weakness, trembling, ataxia, dilated pupils, weakened and slow heartbeat, drowsiness, nausea, vomiting, coldness in extremities or the entire body, labored respiration, paralysis, asphyxia, coma; death is due to respiratory failure. In animals the symptoms usually begin in the hind or lower extremities. The feces may be bloody and accompanied by gastrointestinal irritation and convulsions. Congestion of the respiratory tract is common. Symptoms occur within an hour after ingestion. Death is not always imminent. Abortion may result in pregnant animals. Milk from cows that have eaten *Conium* has an offensive flavor. **Postmortem: gross lesions:** widespread, passive congestion of lungs, liver, and nutrient myocardial vessels; **histological lesions:** cattle show severe mucoid or hemorrhagic enteritis.

**POISONOUS PRINCIPLES:** Gamma-coniceine, coniine, N-methylconiine, conhydrine, lambda-coniceine, and pesudoconhydrine. Toxicity levels vary with the stage of growth (time of year), plant part, and the plant's geographic location. The *Conium* alkaloids are similar in structure and function to nicotine. Gamma-coniceine appears to be the major alkaloid in the vegetative stage. Flowers and immature fruit contain coniine and N-methylconiine. In mature fruit the alkaloid is N-methylconiine. The root contains the least amount of toxins; mature seeds contain the greatest. It has been shown experimentally that the toxic principles in a plant vary even from hour to hour.

**CONFUSED TAXA:** Numerous members of the Umbelliferae superficially resemble *Conium*. Occasionally water-hemlock (see *Cicuta*) is confused with it. *Cicuta* has leaves organized into distinct and separate leaflets of uniform shape, often more than 2 cm wide. *Conium* has dissected leaves with the divisions under 1 cm wide. *Conium* can be confused with *Daucus carota* L. (Queen-Anne's lace, wild carrot), which has distinctly hairy stems, petioles and leaves; poison hemlock stems are glabrous.

**SPECIES OF ANIMALS AFFECTED:** Humans and all species of livestock.

**TREATMENT:** (11a)(b); (20); (6); (2); short-acting barbiturates.

**OF INTEREST:** The odor of the plant, described as "mousey," may be detected on the breath and urine of animals that have eaten the plant. The drug coniine hydrobromide, derived from this plant, is used as an antispasmodic.

# Convallaria

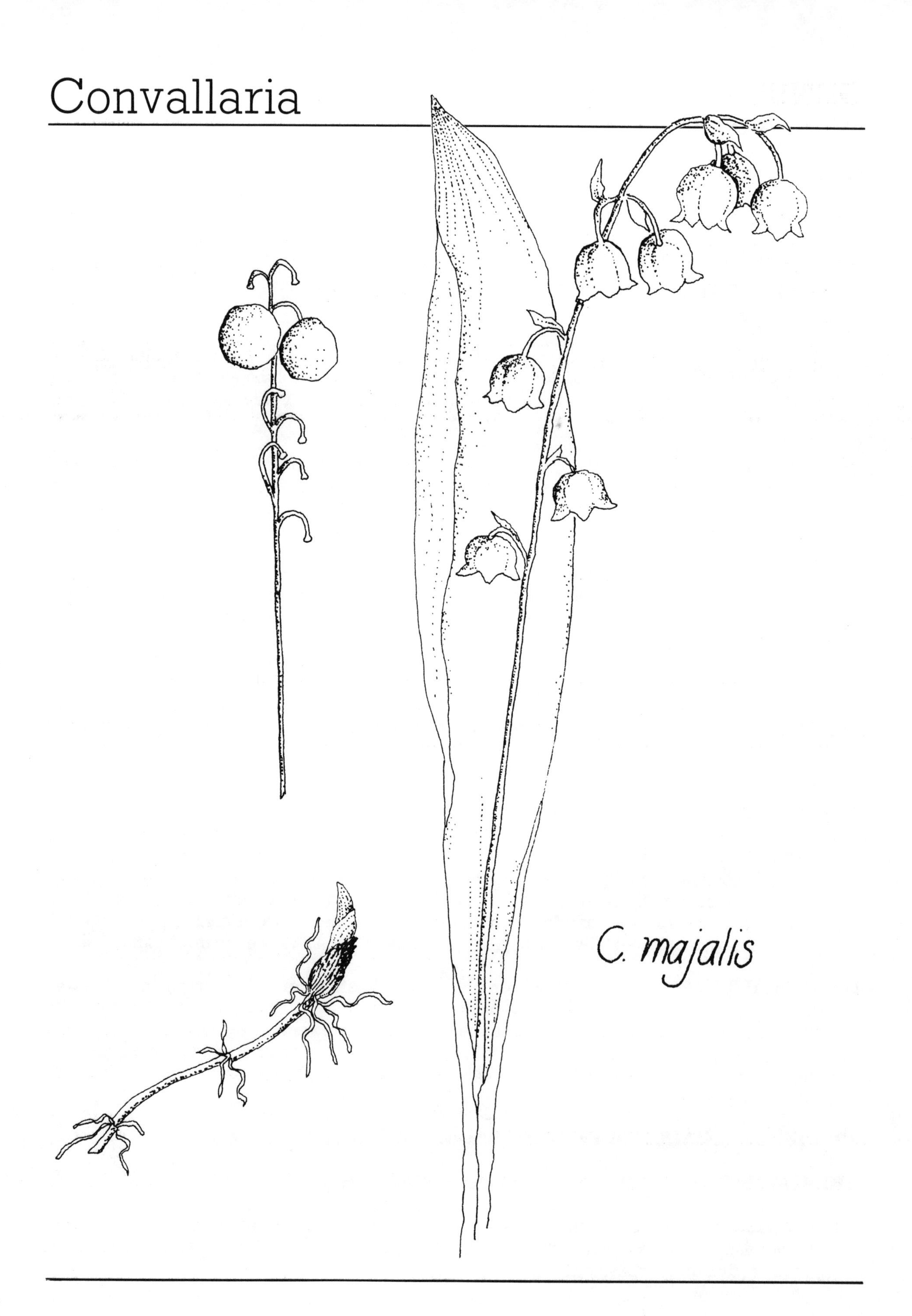

# GENUS: *Convallaria*

*Convallaria majalis* L. — Lily-of-the-valley

---

**FAMILY:** Liliaceae — the Lily Family (see *Amianthium*)

**PHENOLOGY:** Lily-of-the-valley flowers in May.

**DISTRIBUTION:** This cultivated plant frequently persists around foundations and is naturalized in some areas.

**PLANT CHARACTERISTICS:** The familiar lily-of-the-valley produces a one-sided flowering stalk of fragrant, white, nodding, bell-shaped **flowers.** The **fruits** ripen into red berries approximately 1 cm in diameter.

**POISONOUS PARTS:** All parts including pips (underground structures), flowers, fruits, and leaves are poisonous.

**SYMPTOMS:** No reliable reports of livestock losses exist in the literature. There is, however, laboratory-confirmed toxicity. The plant has digitalislike action (see *Digitalis*), including heart arrhythmia (irregular heartbeat) and gastrointestinal upset.

**POISONOUS PRINCIPLES:** More than 20 cardiac glycosides, including convallarin and convallamarin, are known to be produced by this plant.

**CONFUSED TAXA:** Plants are distinct enough not to be readily misidentified.

**SPECIES OF ANIMALS AFFECTED:** Potentially all; livestock and humans are susceptible to the toxins; a report of toxicity to fowl is undocumented.

**TREATMENT:** (11a)(b); (19); (26)

**OF INTEREST:** The drug, convallatoxin, from the blossoms, is used as a cardiotonic. The dried rhizome, known as *Convallaria* root, has also been used as a cardiotonic and diuretic. In veterinary science it has been used as a diuretic and cardiac stimulant.

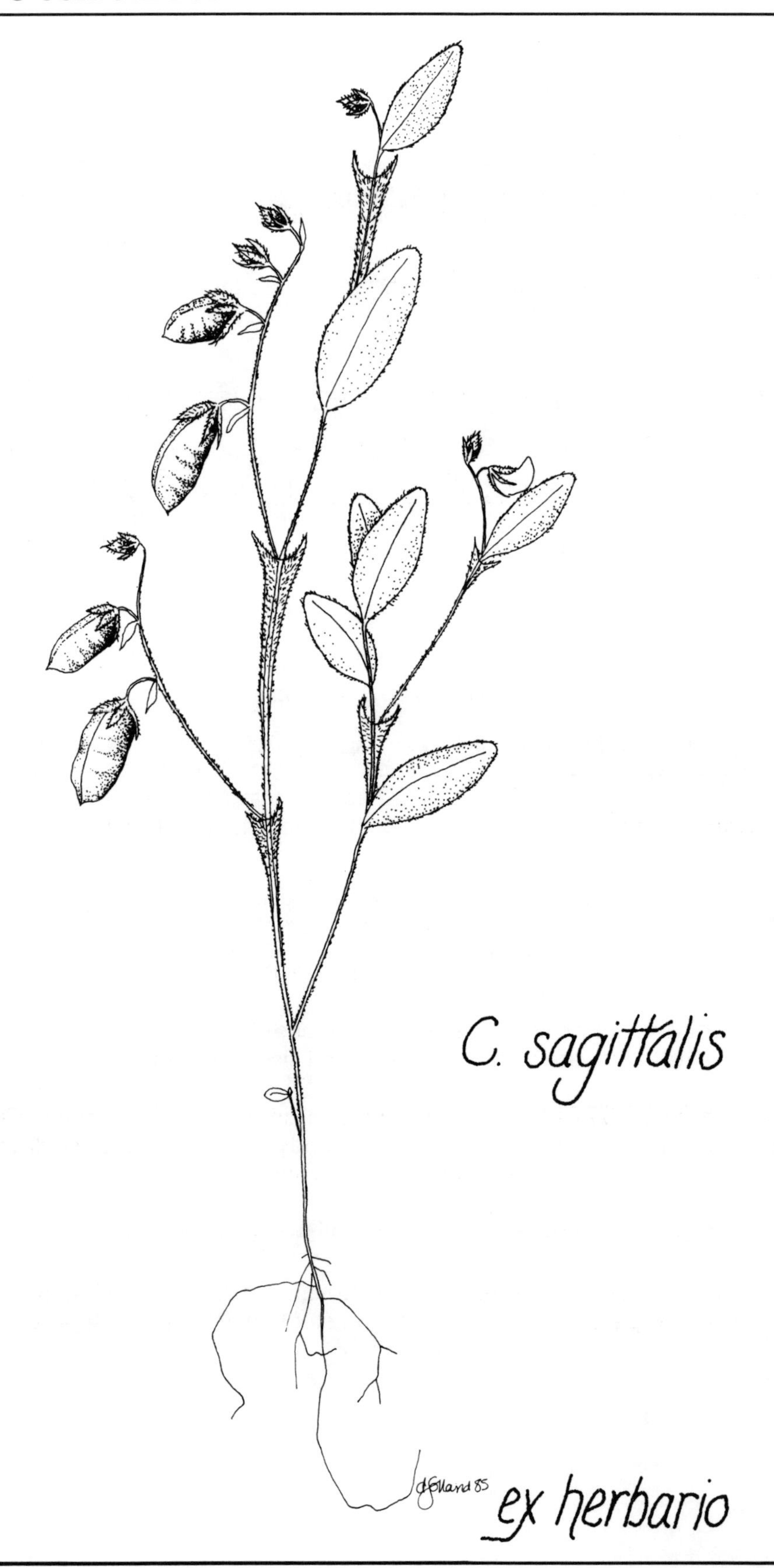
C. sagittalis
ex herbario

# GENUS: *Crotalaria*

*Crotalaria sagittalis* L. — Rattlebox

**FAMILY:** Fabaceae (Leguminosae) — the Bean Family

Legume plants are easily recognized by the familiar flowers, typified by garden beans and peas. The **flowers** are irregular and 5-merous; **calyx:** if prolonged into a tube, often irregular; **corolla:** consisting of 5 petals, the upper **(standard)** exterior and generally larger than the others; the lateral 2 petals **(wings)** are exterior to the 2 lowest petals **(keels),** which enclose the stamens and style; **stamens:** typically 10, their filaments all fused or 9 fused and one free; **fruit:** a 1-celled pod, dehiscent along both sutures, characteristic of the family.

**PHENOLOGY:** Rattlebox flowers throughout an extended period from June to September.

**DISTRIBUTION:** Occurs on dry open soil, waste places, and dry forest clearings.

**PLANT CHARACTERISTICS:** *Crotalaria sagittalis* is a small plant growing to less than half a meter tall, with spreading hairs; **leaf stipules:** decurrent on the stem; **leaves:** simple, entire, sessile, lanceolate lower on the stem to linear toward the top, 3-8 cm, to 1.5 cm wide; **inflorescence:** 2-4 flowered racemes; **flowers:** yellow standard, 8 mm; **stamens:** 10, filaments fused; **fruits:** oblong, sessile pods, 2-3 cm, very inflated, when dry the seeds rattling in the pods; **seeds:** flat, kidney-shaped, brown beans, 2.5 mm long.

**POISONOUS PARTS:** The herbage and seeds are considered toxic. Monocrotaline is present in the entire plant.

**SYMPTOMS:** Livestock show signs of stupor, labored breathing, weakness, emaciation, paralysis, and death. **Postmortem: gross lesions:** hemorrhage, petechiae, or large ecchymoses; organ congestion; abomasum, omasum, and gallbladder are edematous; cirrhosis of liver in prolonged cases; **histological lesions:** pulmonary changes, including emphysema, alternate with atelectasis and hemorrhage.

**POISONOUS PRINCIPLES:** The toxin is probably the pyrrolizidine alkaloid monocrotaline. The additional alkaloids, fulvine and cristpatine, have been isolated and identified as macrocyclic esters of retorsine, which is also a toxic factor in the composite genus *Senecio* (see *Arctium*).

**CONFUSED TAXA:** Lupines (*Lupinus* spp.) resemble rattlebox. In lupine the fruit is flattened rather than inflated, and the leaves are palmately compound instead of simple (see *Lupinus*).

**SPECIES OF ANIMALS AFFECTED:** Generally horses pastured on land containing *Crotalaria sagittalis* will show symptoms. One percent of the animal's body weight fed over two days causes death. In one study cattle fed on hay toxic to horses were not affected.

**TREATMENT:** (11a)(b); (26); possibly treatment with crystalline methionine.

**OF INTEREST:** *Crotalaria sagittalis* is more commonly encountered in the southeastern quarter of the Commonwealth. Related species require warmer climates than Pennsylvania provides.

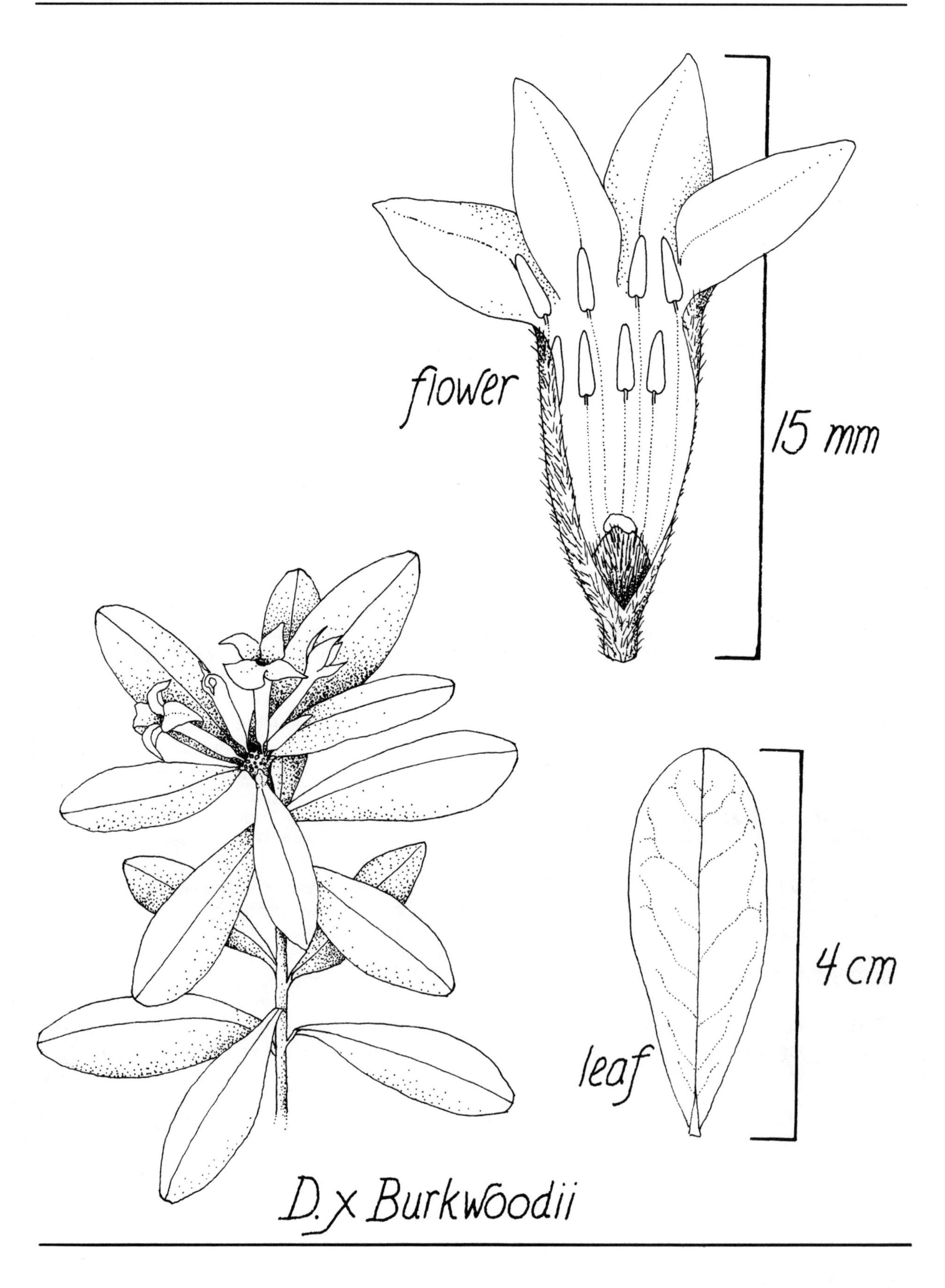
flower
15 mm
4 cm
leaf
D. x Burkwoodii

# GENUS: *Daphne*

*Daphne* x *Burkwoodii* Turril 'Somerset' — Daphne
*Daphne mezereum* L. — Mezereum

**FAMILY:** Thymelaeaceae — the Mezereum Family

Two genera of the mezereum family occur in Pennsylvania, with *Daphne* being a poisonous member. Uncommon in the Commonwealth, it occasionally is grown as an ornamental, either in landscaping or under glass. Some members of the genus are evergreen (southern states); a few are deciduous and cold-hardy. Nurseries and garden centers in Pennsylvania sell *Daphne* x *Burkwoodii* 'Somerset'. It is not known whether this species is poisonous. The data provided below have accumulated from research and case studies of mezereum *(D. mezereum)*. This species, one of the finest and easily cultivated of the daphnes, may be available in Pennsylvania nurseries.

**PHENOLOGY:** Somerset daphne, *D.* x *Burkwoodii* flowers in mid-May, whereas *D. mezereum* produces blossoms in mid-March, April, or early May.

**DISTRIBUTION:** Somerset daphne is a cultivated landscape plant; mezereum is cultivated but has escaped to roadsides, thickets, and old lime quarries in the New England states. It is not known as an escaped plant in Pennsylvania.

**PLANT CHARACTERISTICS:** *Daphne* x *Burkwoodii* is a cross between *D. caucasia* Pall. and *D. Cneorum* L. The flowers are sweetly fragrant, freely produced, and creamy white to pinkish tinged. It is a **shrub,** slightly taller than one meter, evergreen or partially so in protected areas, and deciduous in harsher ones. **Inflorescence:** crowded with 6-16 flowers in terminal clusters, 5 cm wide surrounded by foliage leaves; **flowers:** 1-1.5 cm wide; **fruits:** red.

*Daphne mezereum* is a deciduous **shrub,** 1-2 m tall; **flowers:** very fragrant, 1-1.5 cm wide, produced from the buds of leafless stems in spring, grouped in 2's or 3's; **fruits:** scarlet-red or yellow, 7-10 mm wide, mature in June.

**POISONOUS PARTS:** All parts arc poisonous, especially the "berries" (drupes).

**SYMPTOMS:** Toxicosis includes local irritation, burning or ulceration of mouth and digestive tract, vomiting, bloody diarrhea, internal bleeding, weakness, coma, and death.

**POISONOUS PRINCIPLES:** The toxin daphnin has been isolated. Drupes contain a glycoside; the aglycone is dihydroxy-coumarin.

**CONFUSED TAXA:** There are about 50 species of deciduous or evergreen shrubs in the genus. Cultivated plants with alternate, simple, entire leaves and cylindrical calyx tube with 4 spreading lobes forming the conspicuous part of the flower probably are *Daphne*.

**SPECIES OF ANIMALS AFFECTED:** Humans develop symptoms upon consumption of small quantities of daphne. Children have died from eating only a few drupes.

**TREATMENT:** (9); (11a)(b); (26)

# Datura

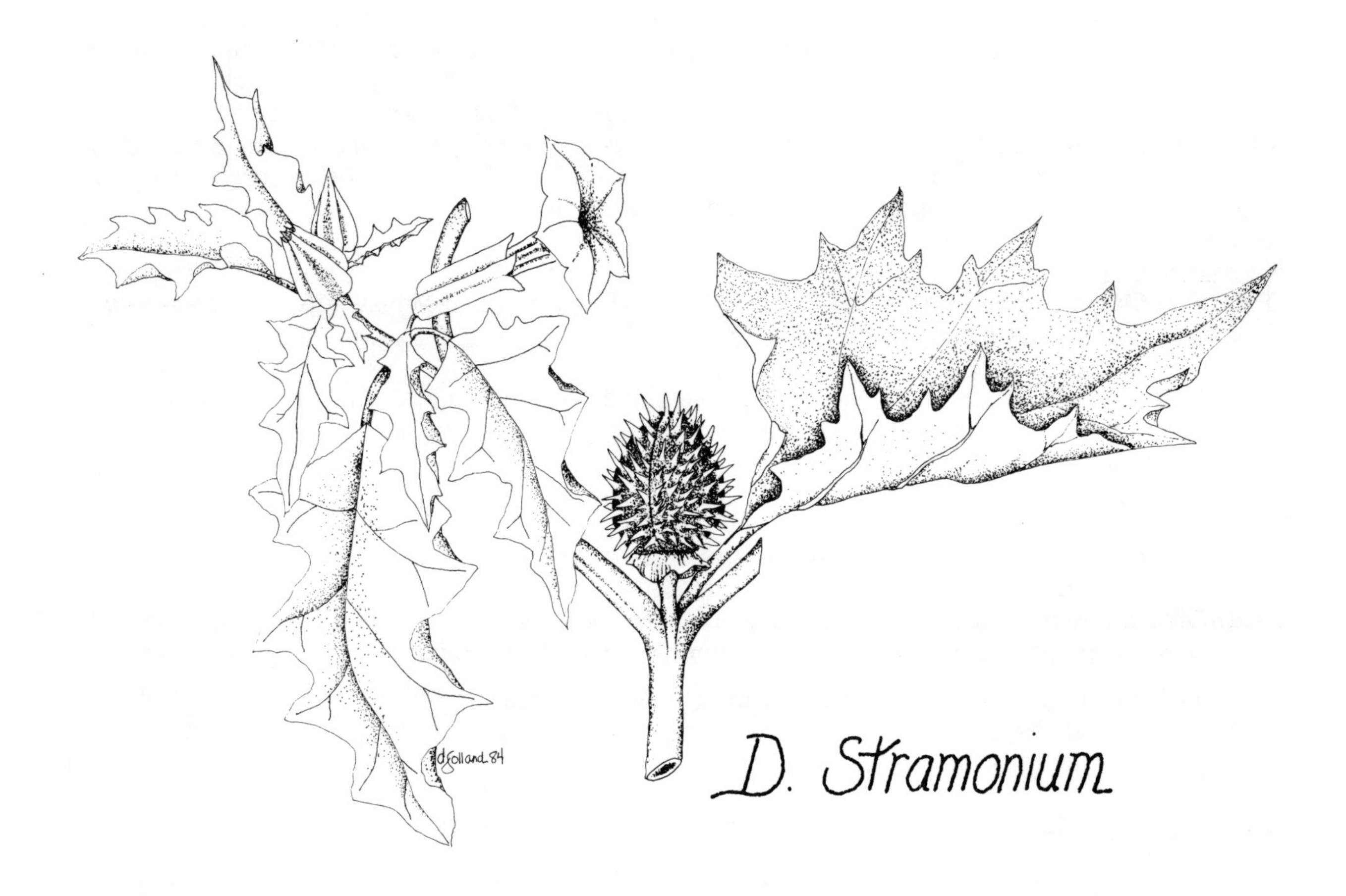


## GENUS: *Datura*

*Datura stramonium* L. — Jimson-weed; moon-lily; thornapple; Jamestown weed

**FAMILY:** Solanaceae — the Nightshade Family

The Solanaceae is a large family of plants with simple, alternate leaves. In our taxa the **flowers** are: bisexual, 5-merous; **calyx:** persistent; **corolla:** rotate, funnelform, or salverform, 5-lobed; **stamens:** 5, borne on the corolla, one or more often appearing different from the rest; **ovary:** superior, mostly 2-celled; **ovules:** many in each cell; **stigma:** 2-lobed; **fruit:** a berry. A distinguishing feature is the **corolla,** which is plicate in bud.

**PHENOLOGY:** *Datura* flowers June through August.

**DISTRIBUTION:** Jimson-weed occurs over the entire state, usually as an inhabitant of dry soil and waste places, dumps, abandoned fields, and in cultivated crops, especially soybeans and corn.

**PLANT CHARACTERISTICS:** This is an annual **plant,** 1.5 m tall with a pungent, "heavy" scent, often branching in two equal forks; **leaves:** 2 x 1.5 dm, with a few teeth; **calyx:** strongly angled in cross section (prismatic) and narrowly 5-winged; **petals:** fused into a tube, white, opening in cloudy weather or evenings, 7-10 cm long; **seed pod:** 3-5 cm, ovoid, with prickles, opening by 4 valves.

**POISONOUS PARTS:** All parts are poisonous, especially seeds and leaves. Lethal dosages for cattle may be 10-14 oz (0.06-0.09% of the animal's body weight). It is estimated that 4-5 g of leaf or seeds would be fatal to a child.

**SYMPTOMS:** In the past several years *Datura* is one plant reported by the U.S. National Clearinghouse for Poison Control Centers as the cause of death. Overdose can occur from excessive ingestion of the herbal medicine Stramonium U.S.P., by accidental poisonings, or intentional ingestion for illicit drug use. Symptoms vary in time of appearance (a few minutes for decoctions to several hours for ingestion of seeds). They include intense thirst, visual disturbance, flushed skin, and central nervous system hyperirritability. Victims become delirious, incoherent, and perform insensible antics. Heart beat may be rapid with elevated temperature. Subjects may be prone to violence, hallucination, convulsions, coma, and death. Ingestion of small amounts produces symptoms; larger amounts, death. Symptoms in livestock approximate those in humans. **Postmortem: gross and histological lesions** are nonspecific.

**POISONOUS PRINCIPLES:** Solanaceous alkaloids (tropane configuration) including atropine, hyoscyamine (isomeric with atropine), and hyoscine (scopolamine). *Datura* alkaloids are useful in medicine. Total content of alkaloids in a plant may be high, varying from 0.25-0.7%. Concentration varies in different parts of the plant, during various stages of development, and under varied growing conditions. The alkaloids are fewer following a rainy period than during clear, dry weather, and concentration decreases during the day but increases at night.

**CONFUSED TAXA:** Two other species, equally poisonous, may be encountered in the Commonwealth. One is *Datura meteloides* DC., with larger flowers (12-20 cm) that open later in the season (July-October). In this species the calyx tube is circular in cross section. Another taxon is *D. metel* L., a European species rarely found in the state.

**SPECIES OF ANIMALS AFFECTED:** Humans, horses, cattle, sheep, hogs, mules, and chickens are susceptible. Human poisonings are more commonly reported. Animals generally avoid the plant, probably because of the intense, pungent smell that is emitted when it is crushed.

**TREATMENT:** (11a)(b); (26); (6); (17)

**OF INTEREST:** American Indians utilized this plant for medicinal and religious purposes. Carlos Castaneda frequently refers to its use in his series of books on the Yaqui Indians (*The Teachings of Don Juan* and others). Soldiers sent in 1676 to quell the Bacon rebellion at Jamestown, Virginia, experienced mass poisoning due to this plant; hence, the common name Jamestown weed.

Because of the hallucinogenic properties of deadly *Datura,* Europeans learned to boil and incorporate plant extracts into fats or oils. These extracts were rubbed on the skin or orifice areas (rectum, vagina) to induce hallucinogenic "flights from reality." Witches during the Middle Ages would anoint a staff (e.g. broom) to apply these compounds; thus, the development of the traditional witch/broom/night flight symbol still seen, especially around Halloween.

# Delphinium

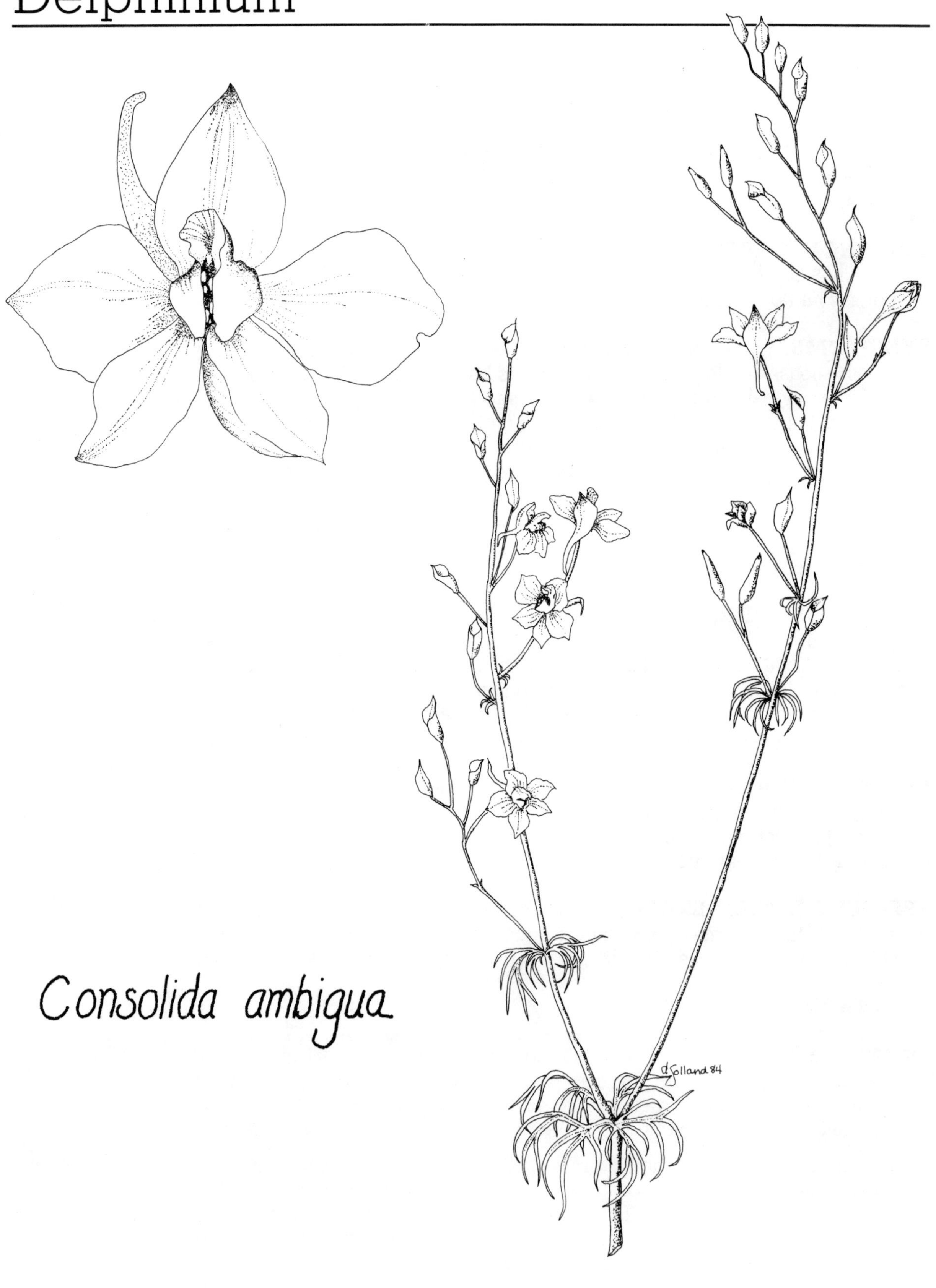

# GENUS: *Delphinium*

*Delphinium* spp. — Larkspur; delphinium

**FAMILY:** Ranunculaceae — the Buttercup Family (see *Actaea*)

**PHENOLOGY:** Two species of Delphinium are encountered in Pennsylvania. *Delphinium exaltatum* Ait. flowers in July and August; *D. tricorne* Michx., in April and May.

**DISTRIBUTION:** *Delphinium* occasionally occurs in rich woods.

**PLANT CHARACTERISTICS:** The genus *Delphinium* has: irregular **flowers; calyx:** resembling a corolla; **sepals:** 5, unequal, blue, purple, or white, the upper one prolonged backward into a **spur; petals:** 4, the upper two inequilateral, each with a long **spur** extending into the spurred sepal, the lower two clawed, abruptly deflexed at the middle; **stamens:** numerous; **pistils:** 1-5; **fruit:** a follicle; **leaves:** palmately cleft.

**POISONOUS PARTS:** The seeds are highly toxic; the young foliage is less poisonous. Toxicity decreases with age of the plant. At flower-bud formation *Delphinium* is half as toxic as in the juvenile stage, and at fruit maturation it is reduced to one-sixteenth. Poisoning in the post-flowering stage, prior to seed formation, is uncommon. Toxicity also varies from species to species.

**SYMPTOMS:** The alkaloids produce digestive disturbance, nervousness, weakness, uneasiness, depression, collapse, muscle spasms, and death by asphyxiation. In larkspur poisoning nausea, bloating, and abdominal pain may be present. **Postmortem: gross lesions:** congestion of internal organs (especially kidneys and superficial vessels); **histological lesions:** acute catarrhal gastroenteritis with diffuse venous congestion.

**POISONOUS PRINCIPLES:** Numerous diterpenoid alkaloids are found in the genus, including delphinine, delphineidine, ajacine, and others. The dried ripe seeds of *Delphinium* contain calcatripine, as well as volatile oil, gum, resin, fixed oil, gallic, and aconitic acids. Poisoning from percutaneous absorption may occur from excessive handling.

**CONFUSED TAXA:** *Delphinium exaltatum* has erect follicles, nontuberous roots, and bifid lower petals. *Delphinium tricorne* is a tuberous-rooted perennial with entire lower petals and divergent follicles.

*Consolida ambigua* (L.) Ball and Heywood (=*Delphinium Ajacis* L.) is an introduced, annual larkspur flowering in the summer. It has one pistil and petals united into one, whereas *Delphinium* has 4 petals. Toxicity and symptoms are identical to *Delphinium*.

*Aconitum* spp. may be confused with *Delphinium* but lack the spurs. *Aconitum* has a solid, pithy stem and short-petioled leaves, whereas larkspurs have hollow stems and long-petioled leaves. *Aconitum* contains the toxin aconitine, which is highly poisonous. Symptoms include violent diarrhea and vomiting, muscular spasms and weakness, respiratory failure, convulsions, and death. Symptoms appear within a few hours after consumption of flowers, leaves, roots, or seeds.

**SPECIES OF ANIMALS AFFECTED:** Humans and most classes of livestock are affected by delphinium alkaloids. Sheep appear much less susceptible than cattle.

**TREATMENT:** (11a)(b); (5-2mg subcutaneously); (27); (6); for poisoned animals subcutaneous injections of physostigmine salicylate, pilocarpine hydrochloride, and strychnine sulfate in the rates 1 grain, 2 grains, and 1/2 grain respectively per 500-600 lb animal.

**OF INTEREST:** Seed extracts are used as a pediculicide in the treatment of head lice.

# Dicentra

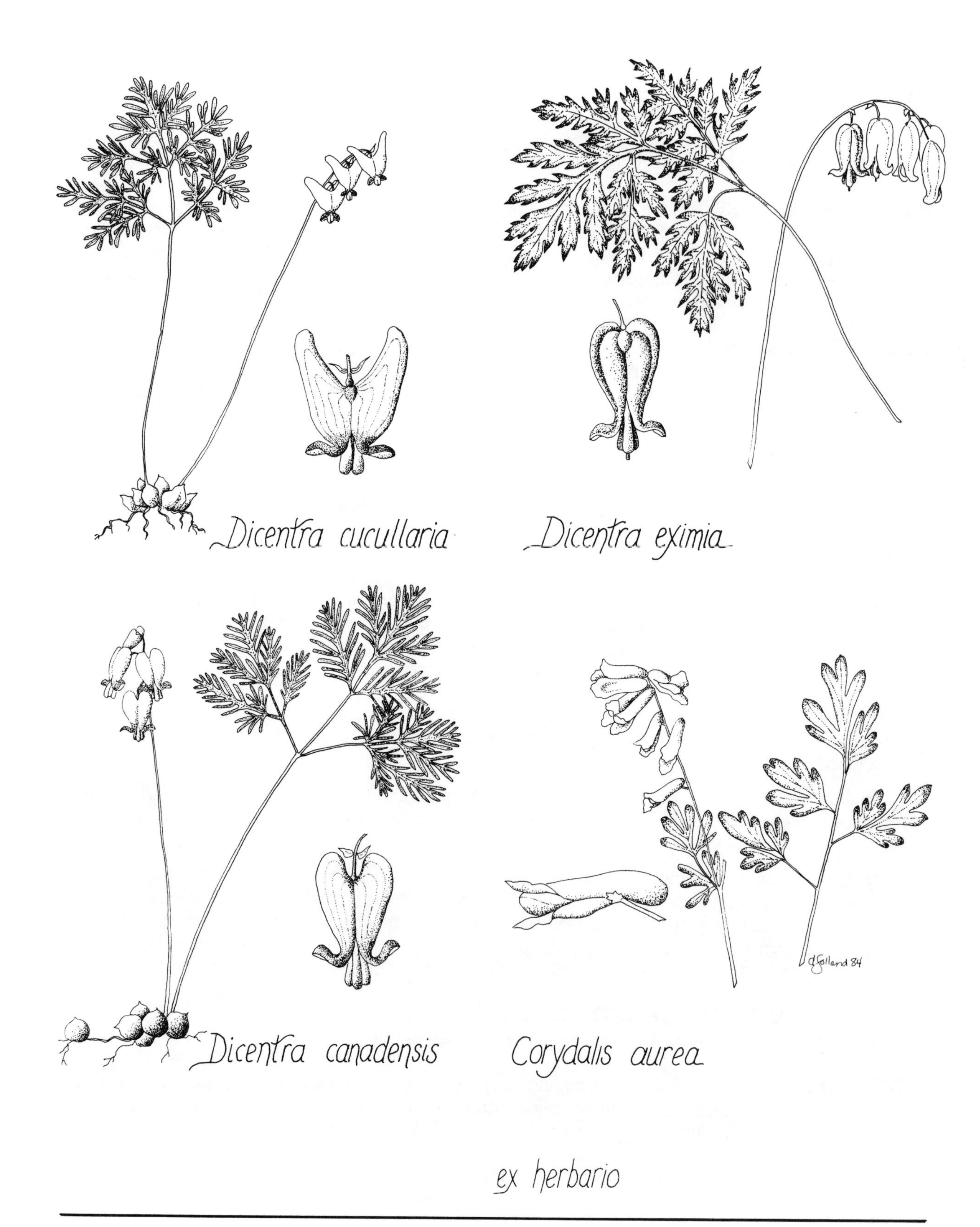
Dicentra cucullaria
Dicentra eximia
Dicentra canadensis
Corydalis aurea
d.Golland 84
ex herbario

# GENUS: *Dicentra*

*Dicentra Cucullaria* (L.) Bernh. — Dutchman's breeches
*Dicentra canadensis* (Goldie) Walp. — Squirrel-corn
*Dicentra eximia* (Ker) Torr. - Bleeding heart

---

**FAMILY:** Fumariaceae — the Fumitory Family

**Flowers** are perfect, 2-nerved, bilaterally symmetrical; **sepals:** 2, falling early from the flower; **petals:** 4, 2 outer and 2 inner; outer 2 petals fused at base, free at the ends, one or both forming **basal sacs;** inner 2 petals slender at base, fused over the stigma at apex; **stamens:** 6; **leaves:** glabrous, herbaceous decompound or dissected; **stems:** watery, juice apparent when crushed.

**PHENOLOGY:** Dutchman's breeches and squirrel-corn flower in spring, usually April to May. Bleeding heart flowers in early summer, June to July.

**DISTRIBUTION:** Dutchman's breeches and squirrel-corn are found throughout the state in rich, moist woods. Bleeding heart can be found in dry or moist woods in the Commonwealth and is an "old time favorite" garden plant.

**PLANT CHARACTERISTICS:** Refer to the family characteristics for a general description of *Dicentra.* Dutchman's breeches and squirrel-corn have flowers in racemes; in bleeding heart the inflorescence is a panicle. In *D. Cucullaria* the outer petal sacs form divergent spurs, whereas in *D. canadensis* the spurs are rounded.

**POISONOUS PARTS:** All parts of these plants are poisonous, especially underground tubers.

**SYMPTOMS:** Only *D. Cucullaria* has been shown (experimentally) to be poisonous, at 2% of the animal's weight. Both aerial and underground portions of the plant produced symptoms within a day although these amounts were not fatal. Behavior of poisoned animals included trembling and running wildly with head held unusually high. Salivation, violent vomiting, and convulsions also were present. Shortly after the onset of symptoms, the animals fell with the head held in the position described, legs rigidly extended; and breathing labored. In one case of experimental feeding the animals could stand, although weakly, within 20 minutes after going down, and recovery was rapid and complete. **Postmortem: gross and histological lesions:** nonspecific.

It may be important to determine the species of *Dicentra* ingested in poison cases since *D. canadensis* failed to elicit symptoms when fed to livestock in amounts equivalent to 2 or 3% of the animal's weight, despite the presence of poisonous alkaloids in the plant.

**POISONOUS PRINCIPLES:** Aporphine, protoberberine, protopine, and related isoquinoline alkaloids are apparently responsible for toxicosis. More than twenty compounds structurally related to poppy alkaloids have been extracted, identified, and named from various species of *Dicentra.* The alkaloids cularine and several of its 0- and N- desmethyl derivatives also are present in leaves and bulbs.

**CONFUSED TAXA:** *Corydalis* is one of the few taxa that can be confused with Dicentra. A close relative whose leaves resemble those of *Dicentra,* it has only one sac per flower. Of the Pennsylvania species of *Corydalis,* both *C. aurea* Willd. and *C. flavula* DC have been suspected of causing livestock loss and should be considered potentially dangerous.

**ANIMAL SPECIES AFFECTED:** Probably all livestock and humans.

**TREATMENT:** (11a) (b); (26)

**OF INTEREST:** These plants have been used in folk medicine to treat a variety of ailments.

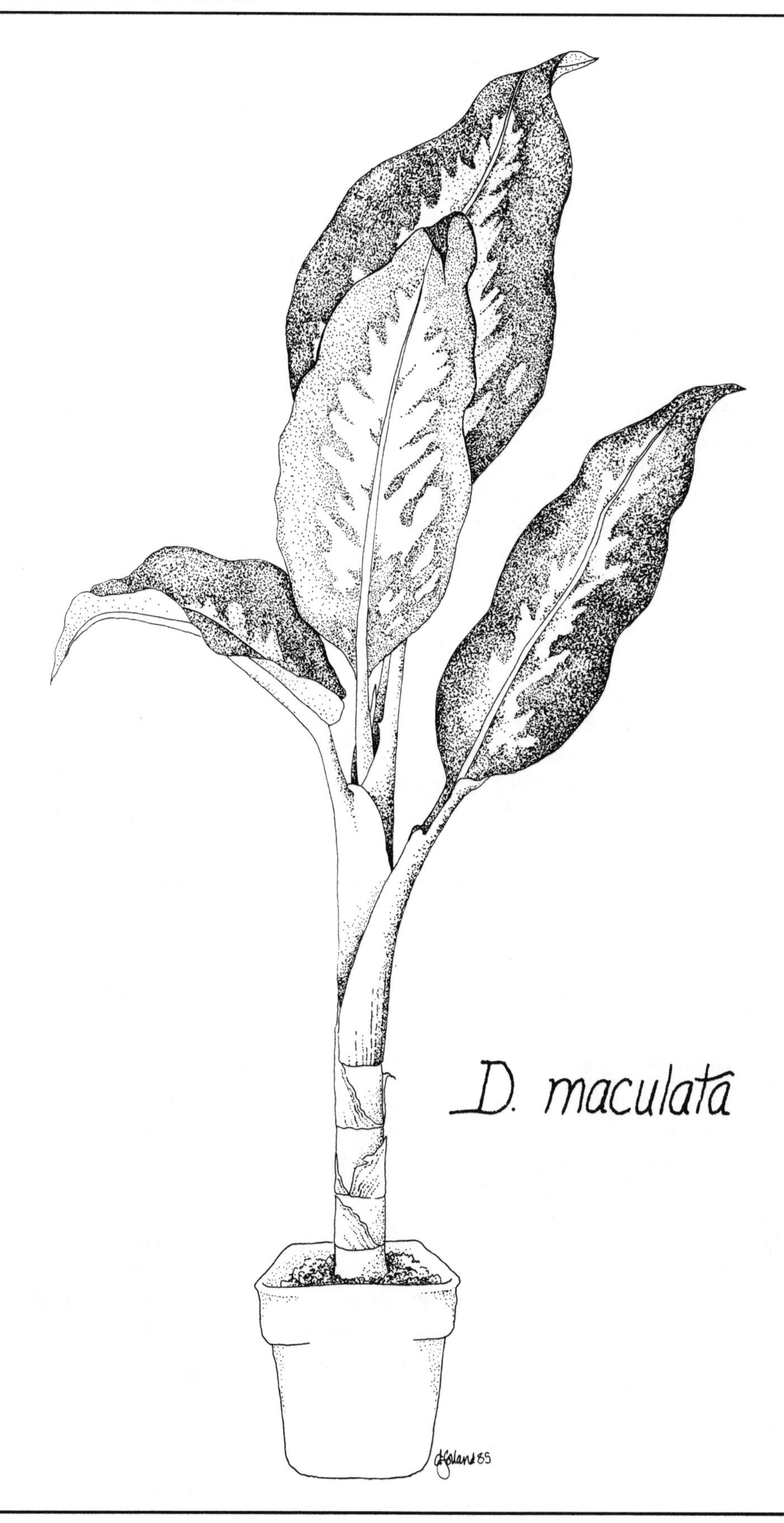
D. maculata

# GENUS: *Dieffenbachia*

*Dieffenbachia* spp. — Dumbcane

**FAMILY:** Araceae — the Arum Family (see *Arisaema*)

**PHENOLOGY:** The flowering requirements of dumbcane are not often met when it is grown as a houseplant.

**DISTRIBUTION:** Native to tropical America, *Dieffenbachia* is now a favorite plant for greenhouses and interior decorations for homes and businesses.

**PLANT CHARACTERISTICS: Stems:** stout, green, girdled with leaf scars, unbranched, bearing leaves toward the top; **leaves:** entire; **petioles:** sheathing; leaves and petioles often spotted or variegated; **flowers:** unisexual; plants sometimes having a skunklike odor when bruised. The **inflorescences** arise from the leaf axils on stalks shorter than the leaf petioles. The inflorescence consists of an erect spike (the **spadix**), which has female flowers at the bottom and male flowers at the top; the two groups of flowers are separated by a short section of naked spadix. A petal-like sheath (the **spathe**) enfolds the lower pistillate flowers and provides a background foil for the staminate flowers.

**POISONOUS PARTS:** All parts contain toxic principles, but the stems are especially poisonous. Oral administration of the toxic component (juice from plants) to guinea pigs showed an $LD_{50}$ between 600 and 900 mg of stem/animal in 24 hr. Injection i.p. produced an $LD_{50}$ of 1 g. A single-dose oral toxicity test in rats showed an $LD_{50}$ over 160 ml/kg of the whole plant juice.

**SYMPTOMS:** Ingestion of dumbcane causes rapid irritation of the mucous membranes, burning, copious salivation due to release of kinins, edematous swelling and thickening of the tongue and lips, local necrosis, and difficulty in swallowing and breathing. The symptoms may last for several days to longer than a week.

**POISONOUS PRINCIPLES:** Mechanical and chemical actions of calcium oxalate may be partly responsible for the reaction to dumbcane. A protein (proteolytic enzyme named "dumbcain") fraction also is possibly responsible, as well as the compound asparagine.

**CONFUSED TAXA:** Some common cultivated house plants such as *Aglaonema* (Chinese evergreen) and *Spathiphyllum* (peace lilies) could be confused with dumbcanes. *Aglaonema* differs from *Dieffenbachia* in having inequilateral leaf bases and the spadix not united with the spathe. In *Spathiphyllum* the leaves are in clusters, not originating on a stout stem as in the above two genera. Two species of *Dieffenbachia* are commonly cultivated: *D. Sequine* Schott. and *D. maculata* (Lodd.) G. Don. Hybridization and natural mutation have created many fancy-leaved forms that are exploited commercially and should be considered poisonous.

**SPECIES OF ANIMALS AFFECTED:** Humans and house pets are susceptible.

**TREATMENT:** (11a) (b); (4); (6); (9); (10); (26); (2 - i.v. diazepam); analgesics for pain (meperidine); intravenous fluids to maintain adequate hydration.

**OF INTEREST:** Other aroids commonly cultivated as house plants should be considered potentially toxic, including *Alocasia* and *Colocasia* (dasheen, elephant ears), *Caladium* (angel-wings), *Anthurium* (tailflower), *Monstera* (Swiss-cheese plant, ceriman), and *Philodendron* (see *Philodendron*). In one case cattle developed severe irritation of mouth and tongue after grazing on *Colocasia*. Cases of severe poisoning from *Alocasia, Caladium,* and *Xanthosoma* (Blue taro, Indian Kale) have not been recorded in the U.S.

# Digitalis

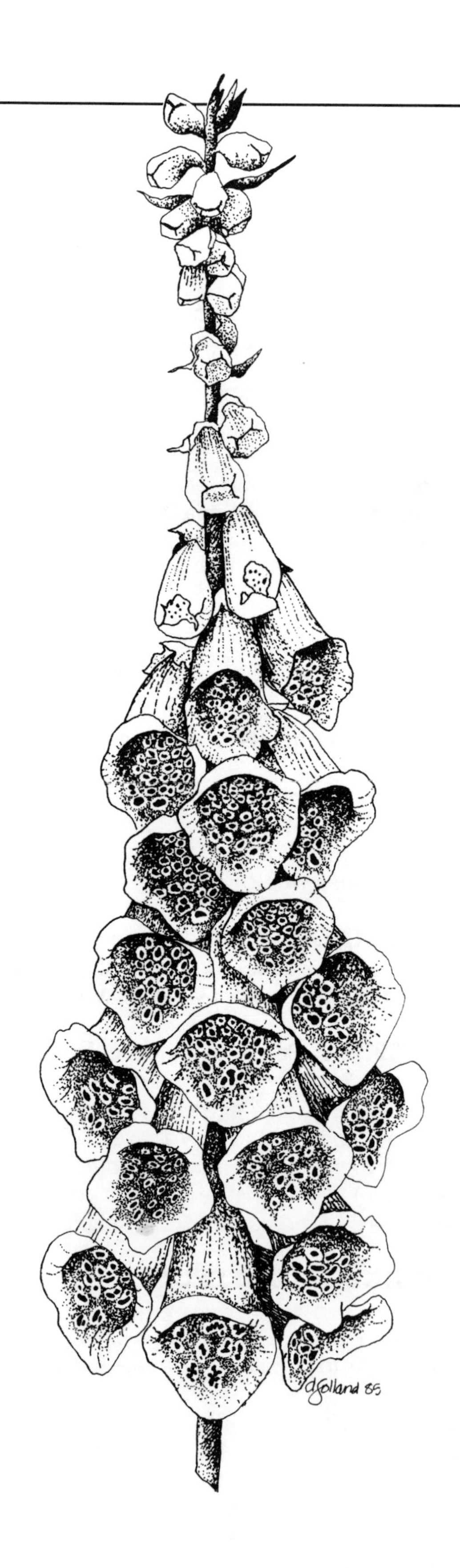

*D. purpurea*

# GENUS: *Digitalis*

*Digitalis purpurea* L. - Foxglove

**FAMILY:** Scrophulariaceae — the Figwort Family

This group, also known as the snapdragon family, is a large, cosmopolitan collection of genera that contains many ornamental plants. Within the family apparently only foxglove is toxic.

**Flowers** are: bisexual, typically irregular, often 2-lipped; **stamens:** 4, or occasionally 2 or 5; **sepals** and **petals:** either 4- or 5-merous; **ovary:** superior, 2-celled; **fruit:** a capsule.

**PHENOLOGY:** Foxglove is a summer-flowering biennial often sold in garden centers already in blossom.

**DISTRIBUTION:** A cultivated plant, foxglove occasionally escapes and is short-lived in the wild in our range. It would be encountered more often in the flower garden.

**PLANT CHARACTERISTICS:** This plant, native to Europe and northwestern Africa, is easily recognized by its showy, terminal, one-sided racemes of large, drooping **flowers.** Other characters include the **basal leaves:** in rosettes, lanceolate to ovate, long-petioled; **stem leaves:** alternate, simple, sessile or nearly so; **calyx:** 5-parted; **corolla:** to 7 cm long, purple to pink or white with conspicuous spots inside; **stamens:** 4.

**POISONOUS PARTS:** The herbage, both fresh and dried, contains powerful, highly toxic compounds.

**SYMPTOMS:** In humans toxic reactions include gastric upset, nausea, diarrhea, abdominal pain, severe headache, pulse and cardiac rhythm abnormalities, mental irregularities, drowsiness, tremors, convulsions, and death. In livestock symptoms are similar and include bloody stools, lack of appetite, and the urge to urinate.

**POISONOUS PRINCIPLES:** Digitalis, an extract of foxglove, has been used medicinally for many years; hence, a large body of information has accumulated concerning it. Physiologically active chemical constituents found in foxglove include digitoxin (0.2-0.4%), digitonin, digitalin, antirhinic acid, digitalosmin, and digitoflavone. The $LD_{50}$ (oral) for digitoxin in guinea pigs is 60 mg/kg of body weight, and in cats is 0.18 mg/kg of body weight. The toxins are cardiac or steroid glycosides. The aglycones are derivatives of cyclopenteno-phenanthrene; and the sugars, unusual methyl pentoses. They influence the heart in two ways: stronger cardiac contractions and slower contractions through stimulation of the vagus, prolonging diastole. *Digitalis* is of much importance to modern medicine.

**CONFUSED TAXA:** No other garden plants produce tall, (to more than 1.5 m) one-sided racemes of purple, spotted flowers.

**SPECIES OF ANIMALS AFFECTED:** Humans, usually from drug overdoses, and livestock, from grazing, have been poisoned by foxglove alkaloids. Poisoning in animals can result from hay contamination in addition to browsing fresh material. Susceptible animals include pigs, cattle, horses, and turkeys.

**TREATMENT:** (11a)(b); (26); (5); (19)

# Equisetum

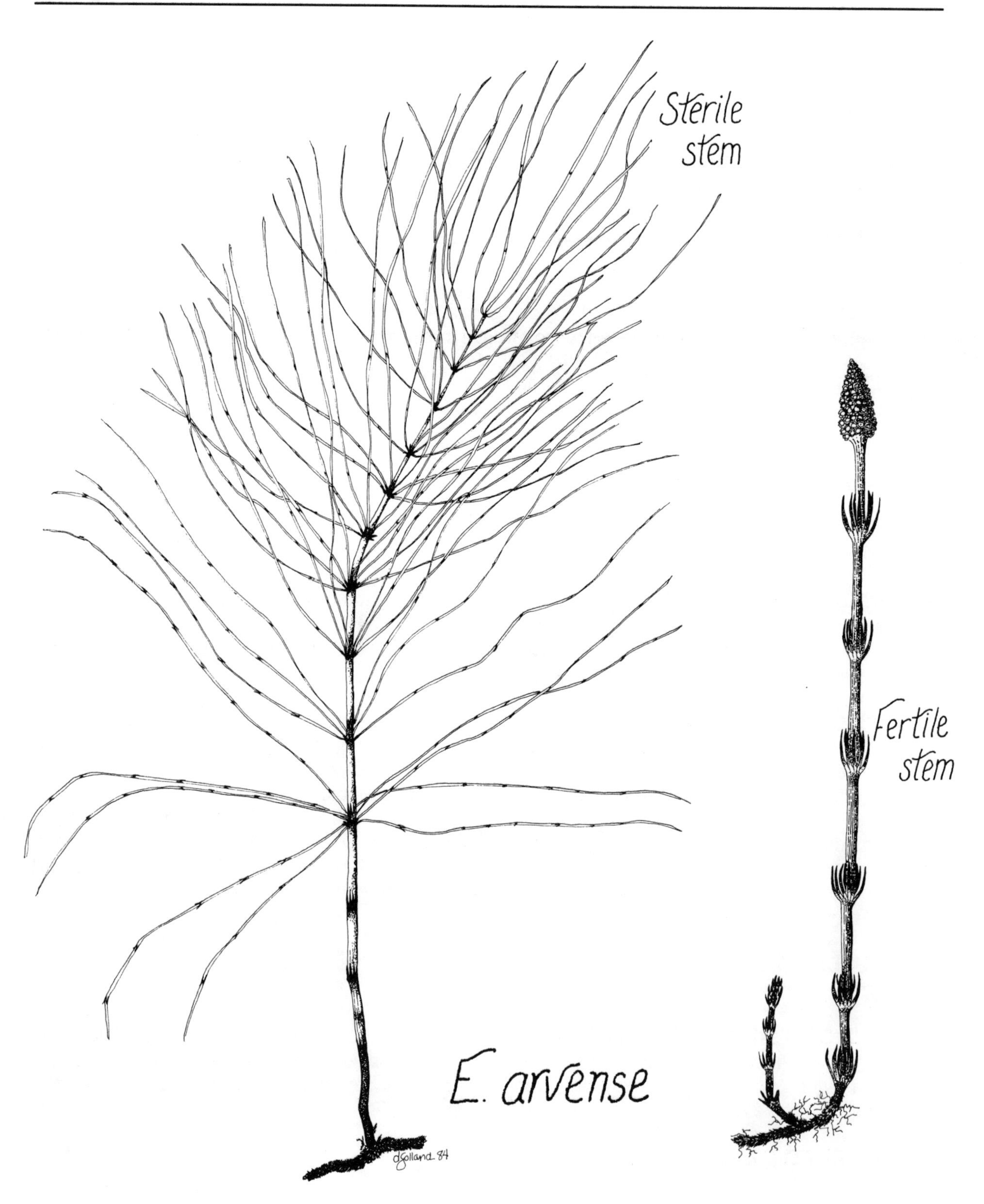


**GENUS:** *Equisetum*
*Equisetum arvense* L. - Common horsetail

**FAMILY:** Equisetaceae — the Horsetail Family

This family of fern allies contains only one genus — *Equisetum*. Plants are primitive, spore-bearing, annual or perennial, rhizomatous herbs having vascular tissue.

**PHENOLOGY:** The fruiting period is April to July.

**DISTRIBUTION:** *Equisetum arvense* thrives in a broad range of habitats from moist to wet, or in moderately dry sandy soil; it may grow in fields, woods, on stream banks, or along roadsides.

**PLANT CHARACTERISTICS:** Horsetail is not a true flowering plant with sepals and petals. The reproductive structure is a **sporophyll** in a terminally, spikelike **cone** composed of shield-shaped spore-bearing structures. Other characteristics include, **stem:** hollow, one large central canal surrounded by smaller ones under each ridge of the stem; jointed; impregnated with silica; striated or grooved; rushlike; **branches:** whorled; **leaves:** marginally united into a sheath around each node. Plants bear nongreen reproductive structures in spring and green vegetation during the remainder of the growing season.

**POISONOUS PARTS:** All parts, green and dried, can be toxic. Hay containing 20% or more of *E. arvense* causes poisoning symptoms in horses in 2-5 weeks.

**SYMPTOMS:** Toxicosis is similar to bracken poisoning (see *Pteridium*). Apetite remains normal until near the end of the illness in *Equisetum* poisoning, whereas it is lost early in bracken poisoning. Ataxia, difficulty in turning, and the body wasting away followed by general weakening are early signs. In later stages animals may become constipated and the muscles rigid. Pulse rate increases and weakens, and the extremities become cold. The cornea of the eye may become opaque. Before death, the animal becomes calm and comatose. If poisoning is discovered early, the toxic plants removed from the diet, and proper nutrition given, animals can recover rapidly.

Horses are not infrequently affected by *E. arvense*. In advanced cases when a horse "goes down" and cannot arise, the animal becomes nervous, making frantic attempts to stand. When a poisoned horse is exercised, it will tremble and become muscularly exhausted. Cattle are not readily affected by *E. arvense*. In experiments with cattle, the only result was marked loss in the general condition of the animal over a forty-day period. **Postmortem: gross lesions:** none evident; **histological lesions:** diffuse lesions in the cerebral cortex such as polioencephalomalacia or cerebrocortical necrosis; bilaterally symmetrical zones of malacia involving various nuclei of the brain.

**POISONOUS PRINCIPLES:** The enzyme thiaminase is responsible for poisoning in nonruminants. The previously suspected silica, aconitic acid, palmitic acid, nicotine, methoxypridine, equisitine, palustrine, and dimenthyl sulfide components are not of themselves toxic enough to produce poisoning. However, they may complicate the toxicosis. The toxic agent for ruminants is unknown and generally not fatal.

**CONFUSED TAXA:** No unrelated plants occurring in Pennsylvania look like species of the distinctive genus *Equisetum*. All members of this common genus should be considered poisonous.

**SPECIES OF ANIMALS AFFECTED:** Horses primarily are affected, with sheep to a lesser degree. Equisetosis is rarely fatal for cattle.

**TREATMENT:** (11a)(b); (26); massive doses of thiamine.

**OF INTEREST:** The stems, which contain silica crystals, are sometimes used to polish metal; hence the name scouring rush. Plants are occasionally used as an ornamental in moist places such as around ponds.

# Eupatorium

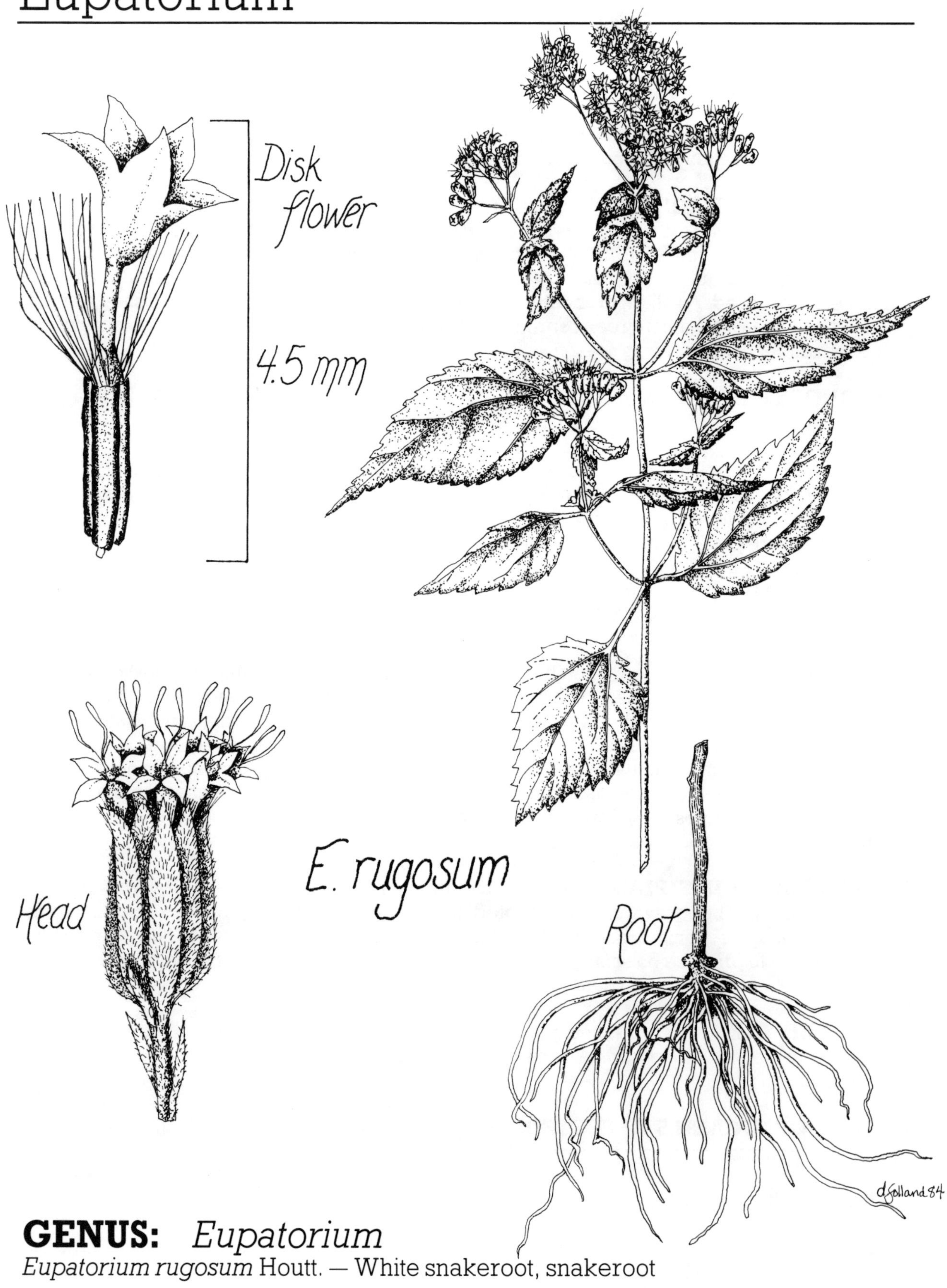


**GENUS:** *Eupatorium*
*Eupatorium rugosum* Houtt. — White snakeroot, snakeroot

**FAMILY:** Compositae (Asteraceae) — the Daisy Family (see *Arctium*)

**PHENOLOGY:** Snakeroot flowers from July through October.

**DISTRIBUTION:** *Eupatorium rugosum* grows in eastern North America. Abundant in Pennsylvania, it is found in moist areas, rich open woods, and along streams. It often forms dense colonies in areas where logging has cleared regions of the forest.

**PLANT CHARACTERISTICS:** The heads of white flowers are small and contain only **disk flowers:** 3 to 4 mm long; **heads:** contain 10 to 30 flowers; **leaves:** thin, opposite, evidently petioled, sharply serrate, acuminate, the larger ones 6 (to 18) x 3 (to 12) cm long; **stems:** 1.5 dm, from a shallow mat of fibrous, perennating roots.

**POISONOUS PARTS:** The poisonous parts are the leaves and stems; toxicity decreases with drying. Seasonal or ecological variation may affect toxic principles; frost does not reduce toxicity.

**SYMPTOMS:** The following conditions are recorded in toxicosis: trembling, especially of the flank and hind legs; slow, lethargic or sluggish behavior ("the slows"); stiffness in movement or ataxia; coma; and death. Horses seem less prone to trembling. Livestock may exhibit constipation, nausea, vomiting, slobbering, loss of appetite, or labored or rapid breathing. An acetone odor on the breath may result from ketosis in severely poisoned animals. **Postmortem: gross and histological lesions:** fatty degenerative changes in liver and kidney; heart and gastrointestinal hemorrhages.

**POISONOUS PRINCIPLES:** Poisoning is possible due to an unstable, fat-soluble alcohol called tremetol, which has a phenyl nucleus and an incompletely characterized resin acid. Tremetol is an aromatic, straw-colored oily liquid. The principal ketone, tremetone, is also suspected to be toxic.

**CONFUSED TAXA:** There are more than twenty species of *Eupatorium* in eastern U.S., with most encountered in Pennsylvania. Because *E. rugosum* may be difficult to identify accurately, any suspected material should be examined by a specialist.

**SPECIES OF ANIMALS AFFECTED:** Species known to be susceptible are fowl, sheep, horses, cattle, hogs, and humans; illness may be experimentally induced in various laboratory animals.

**TREATMENT:** (26); treat for anuria, liver, and kidney damage.

**OF INTEREST:** "Trembles" and "the slows" are not the only illness produced from *Eupatorium* toxicosis. "Milksickness" was a common disease in the Colonial period, reaching its peak in the first half of the 1800's. Since tremetol is readily excreted in milk, poisoning can occur from ingestion of milk from lactating animals that have eaten *Eupatorium*. During the early pioneer period, entire villages were abandoned and human population centers were reduced to less than one-half the original number due to milksickness because the etiology of the disease remained a mystery. The amount of plant consumed to elicit sickness varies from 1-20% of the animal's weight and depends on many factors. Because the toxin is cumulative, the onset of symptoms varies from less than 2 days to as much as 3 weeks. Death follows from 1 day to 3 weeks after the symptoms appear. Recovery is good if the disease can be countered before ketosis appears. In many cases, prognosis is poor and recovery is rare, slow, and incomplete. Milksickness in humans begins as a few days of weakness, loss of appetite, abdominal pain, and violent vomiting followed by obstinate constipation, severe thirst, loss of ingested fluids by vomiting, tremors, acetone breath, prostration, delirium, coma, and death. The disease can be fever-producing or the temperature may be subnormal. Mortality ranges from 10-25%, but the massive loss of human life seen in centuries past does not occur owing to current practices of animal husbandry and the pooling of milk from many producers.

# Euphorbia

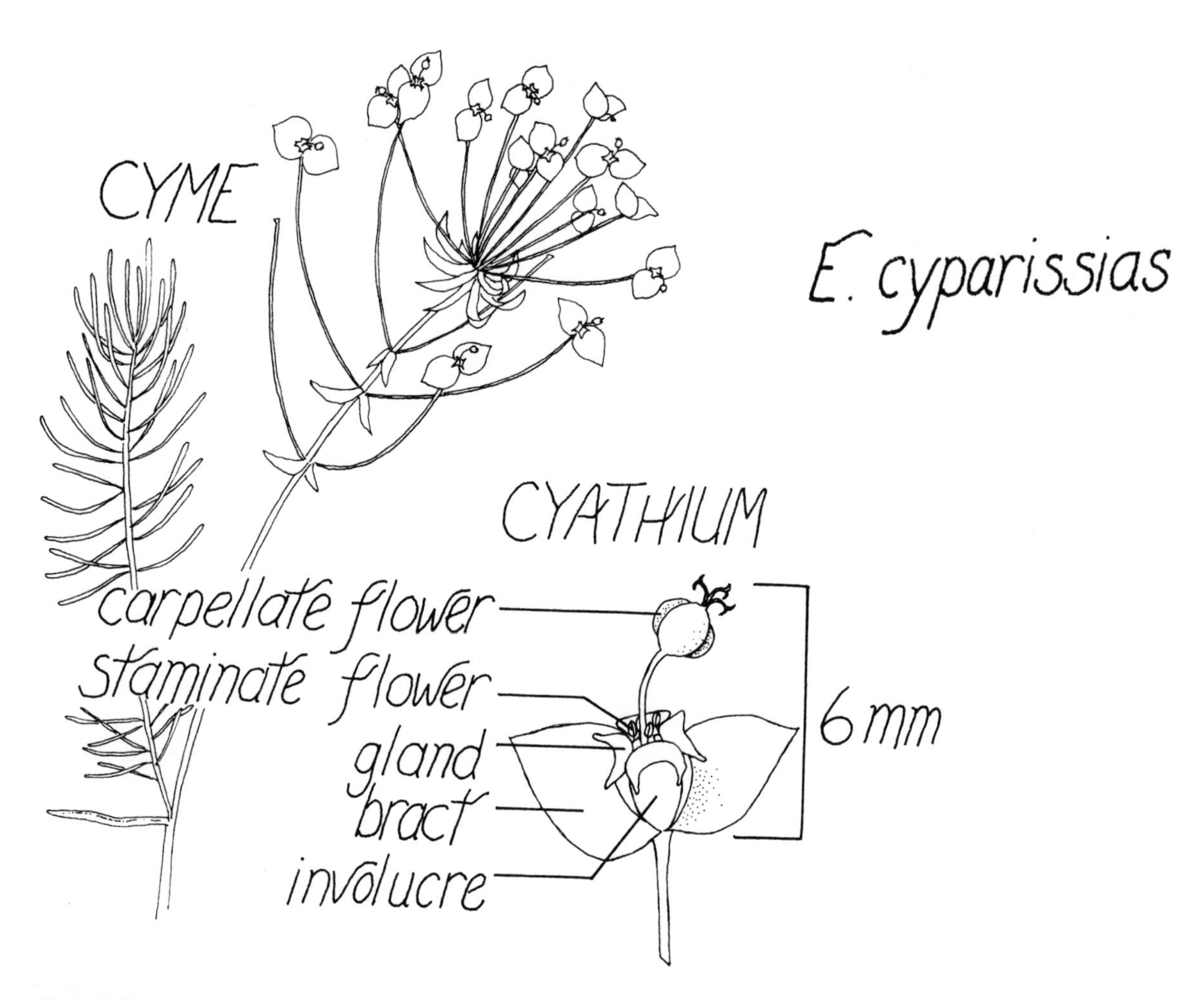


## GENUS: *Euphorbia*

*Euphorbia cyparissias* L. — Cypress-spurge
*Euphorbia maculata* L. — Wartweed
*Euphorbia marginata* Pursh. — Snow-on-the-mountain

**FAMILY:** Euphorbiaceae — the Spurge Family (see *Codiaeum*)

**PHENOLOGY:** *Euphorbia* spp. bloom from May to September depending on the taxon.

**DISTRIBUTION:** More than twenty species of *Euphorbia* are found in Pennsylvania. They occur as weeds (*Euphorbia maculata* L.), as garden plants (*E. marginata* Pursh.), or as plants escaped from cultivation (*E. cyparissias* L.).

**PLANT CHARACTERISTICS:** The *Euphorbia* cyathium contains **staminate flowers:** several, each containing one **stamen; pistillate flower:** one, containing one **pistil; ovary:** 3-celled, 3-ovuled; **styles:** 3; **involucre:** 4- to 5-lobed, usually bearing **glands** in the sinuses. Our plants are **herbs,** usually with milky, acrid **juice.**

**POISONOUS PARTS:** The whole plant, fresh or dried, is poisonous.

**SYMPTOMS:** Diarrhea, collapse, and death may result from ingestion of *Euphorbia cyparissias.* Hay contaminated with *E. cyparissias* has caused death in cattle.

*Euphorbia maculata,* growing predominantly in a pasture, caused a 30% loss of Hampshire lambs. *Euphorbia maculata* fed to lambs at a rate of 0.62% body weight caused death within hours. Surviving lambs were photosensitized so that exposure to sunlight produced edematous enlargement of the head. Toxicity may increase during July and August when rains follow a period of drought.

*Euphorbia marginata* produced diarrhea and emaciation, lasting several months, when 100 oz was fed to cattle. Consumption of this plant can cause blistering, irritation, and inflammation of the upper digestive tract. The sap has been noted to cause contact dermatitis on the legs and face of horses. The death of a young woman, who drank a decoction of *E. marginata* in an effort to abort, is reported.

**POISONOUS PRINCIPLES:** Irritant cocarcinogenic diterpenoids have been isolated as toxins in some *Euphorbia* species.

**CONFUSED TAXA:** *Euphorbia cyparissias* is a colonial plant from creeping rootstocks; leaves: linear, entire, crowded, 3 cm long, 1-nerved; cyathia: umbel-like cymes, yellowish green when young, becoming purplish red; glands: yellow, crescent shaped with 2 short horns.

*Euphorbia maculata* are prostrate plants growing like mats over the ground; leaves: all opposite, oblique at base, dark green, often with a median red spot; stipules: present; glands and petaloid appendages: 4. Botanists are not in full agreement as to the identity of plants to be included in this species. Some plants that may prove to be equivalent to *E. maculata* L., are *E. supina, E. chamaesyce* and *E. hirta.*

*Euphorbia marginata* grows to 2 m; leaves: alternate, sessile, broadly ovate to elliptic; leaves subtending the inflorescence: whorled; marginated with white, or entirely white; involucral lobes: fringed; appendages: 5, white, conspicuous; fruit: 3-lobed, 6-7 mm in diameter.

**TREATMENT:** (11a)(b); (26)

**OF INTEREST:** All species of *Euphorbia* can be expected to contain the complex esters believed responsible for poisonings and are probably capable of eliciting an allergic reaction. Reported deaths from *Euphorbia* poisoning are rare but livestock can be seriously affected.

*Euphorbia corollata* L. (flowering spurge) has been implicated in the poisoning of livestock. This widely distributed weed flowers from May though September and has 5 cyathial glands and white petaloid appendages. It has been used in small doses for diaphoresis and as an expectorant.

*Euphorbia heterophylla* L. (fire-on-the-mountain), commonly cultivated in the Midwest, is now becoming prevalent as an escaped plant in the East. Diagnostic characters include the absence of petaloid appendages on the singular gland of the involucre and stem leaves that are mostly alternate, glabrous above, and hairy beneath. A closely related species, *E. dentata* Michx., has opposite leaves with hairs on both sides. It too is found in increasing numbers on dry sites, along roadsides, and in waste places. Both are to be treated with suspicion, although no records indicate poisonings from these taxa.

Other species to be alerted to are *E. Preslii* Guss., *E. esula* L. (leafy spurge), and *E. Peplus* L. (petty spurge). The last-named species is locally abundant in vegetable gardens, and unconfirmed reports indicate that dogs eating a low concentrated mixture of it and grass develop violent diarrhea; it has proven lethal to human beings. *Euphorbia lathyris* (caper-spurge) was used as a medicinal folk remedy. This toxic species can still be found around old homesteads and as an escaped plant.

*Euphorbia* commonly grown as houseplants include *Euphorbia Milii* Ch. des Moulins (crown-of-thorns) and *E. pulcherrina* Willd. (Poinsettia). The crown-of-thorns is a woody, spiny plant with cyathia subtended by bright red bracts. When bruised, it produces an irritant, white, milky latex. No cases of severe poisoning have been reported in the literature, but crown-of-thorns may be toxic if consumed in quantity. Poinsettias are popular Christmastime house plants. No well-documented cases of severe poisoning exist despite many cases of fruit, bud, and leaf ingestion. The irritant milk may produce symptoms of gastroenteritis including abdominal pain, vomiting, and diarrhea.

# Glecoma

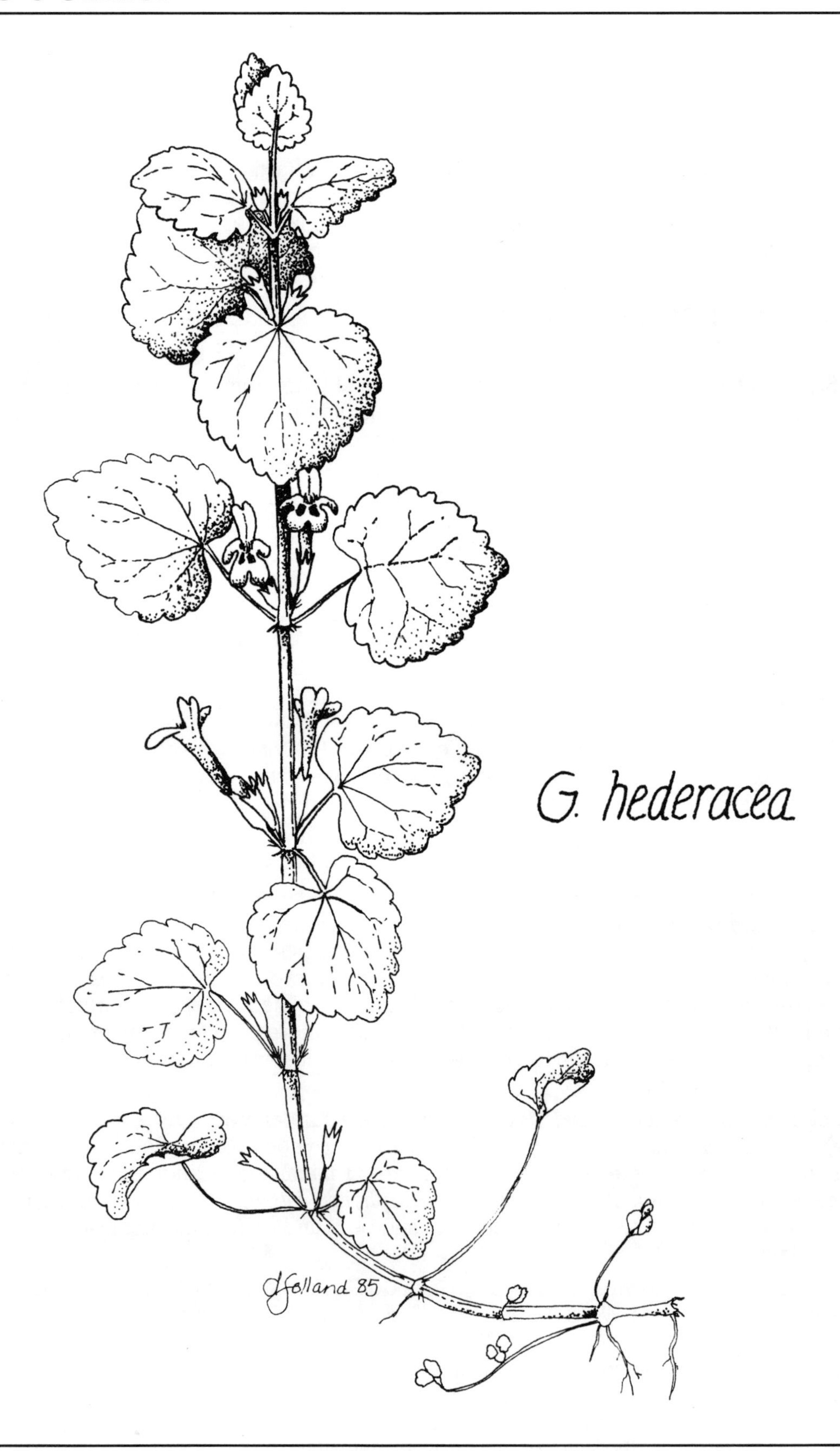

# GENUS: *Glecoma*

*Glecoma hederacea* L. — Ground-ivy; gill-over-the-ground; creeping Charlie; run-away-robin

---

**FAMILY:** Labiatae (Lamiaceae) — the Mint Family

This is a large group of plants known for glands that secrete pungent, volatile oils that may be toxic in large amounts. Many plants in this family are cultivated as ornamentals or as sweet herbs. Characteristic features for this family include **stems:** square in cross section; **leaves:** opposite, 4-ranked, simple, without stipules, glandular; **flowers:** irregular; **calyx:** 4- to 5-lobed, 2-lipped, persistent; **corolla:** 4- to 6-lobed, 2-lipped, petals conspicuously united; **stamens:** 4 (rarely 2), inserted on the corolla; **ovary:** superior, deeply lobed; **style:** 1 from the center of the ovary lobes; **fruit:** 4 one-seeded nutlets.

**PHENOLOGY:** Ground- ivy flowers April through June.

**DISTRIBUTION:** Found in moist fields or woods or in disturbed soil, including roadsides and yards.

**PLANT CHARACTERISTICS:** Plants are prostrate or creeping; **leaves:** cauline, petiolate, 1-3 cm, crenate rotund-cordate to cordate-reniform; **stem:** retrorsely scabrous to subglabrous; pilose at the nodes; **flowers:** 13-23 mm across, on short pedicels; **petals:** blue violet, marked with purple spots; usually 3 flowers per axil.

**POISONOUS PARTS:** All parts are toxic in green or dried condition.

**SYMPTOMS:** Salivation, sweating, dyspnea, panting, dilated pupils, anxious look, cyanosis, and possibly pulmonary edema can be manifested. **Postmortem:** pulmonary edema and cerebral hyperaemia.

**POISONOUS PRINCIPLES:** Physiologically active volatile oils are responsible for toxicosis.

**CONFUSED TAXA:** *Glecoma* is sometimes mistaken for *Lamium* (dead nettle), but the flowers are distinctly pediceled (in loose cymules) in *Glecoma* and sessile (in dense cymules) in *Lamium*.

**SPECIES OF ANIMALS AFFECTED:** Apparently only horses are susceptible to *Glecoma* toxins.

**TREATMENT:** (11a)(b); (26)

**OF INTEREST:** *Lamium amplexicaule* L., commonly encountered as an element of our spring flora, causes "staggers" in sheep, horses, and cattle. *Stachys arvensis* L., fieldnettle, is responsible for nervous disorders in livestock, especially sheep. It too is a Pennsylvania resident.

# Gymnocladus

**GENUS:** *Gymnocladus*
*Gymnocladus dioica* (L.) C. Koch — Kentucky coffee-tree

**FAMILY:** Caesalpiniaceae — the Caesalpina Family

The bean family, Leguminosae, is naturally divided into three subfamilies: the Faboideae, the Mimosoideae, and the Caesalpinioideae. Some authorities classify the three subfamilies as separate families as they are presented here. The Fabaceae family description is found under the *Crotalaria* entry. No members of the Mimosaceae found within Pennsylvania are toxic. The Caesalpiniaceae are characterized by **flowers:** perfect or unisexual, regular or sometimes irregular; **hypanthium:** well developed, often irregular; **sepals:** 5; **petals:** 1-5; **stamens:** typically twice as many as the sepals; **ovary:** 1; **fruit:** a legume splitting along 2 sutures.

**PHENOLOGY:** *Gymnocladus dioica* flowers in May.

**DISTRIBUTION:** Kentucky coffee-tree is found in rich moist woods. It is seldom abundant, frequently occurring as single trees.

**PLANT CHARACTERISTICS:** Kentucky coffee-tree is a tall **tree,** to 30 m; **plants:** bear on one tree flowers partly perfect and partly pistillate, and on another tree flowers partly perfect and partly staminate; **flowers:** regular, perfect or unisexual, 5-merous; **hypanthium:** tubular, 10-15 mm; sepals and petals in a single series, 8-10 mm; **stamens:** 10, distinct, alternately long and short; **fruit:** red-brown, woody, flat, thick, often exuding a yellow resin when broken; **seeds:** few per pod, large, separated by pulp; **branches:** stout, without small twigs or thorns; **leaves:** very large, bipinnately compound, new leaves often pink; **leaflets:** 2-3 cm wide; **inflorescence:** terminal panicles of greenish-white flowers.

**POISONOUS PARTS:** The sprouts, foliage, and fruit are poisonous.

**SYMPTOMS:** Severe gastrointestinal irritation and narcoticlike effects on the nervous system have been reported. **Postmortem: gross and histological lesions:** congestion of the mucous membranes.

**POISONOUS PRINCIPLES:** The cause of toxicosis is unknown but may be due to the presence of alkaloids. Cytisine, a toxic quinolizidine alkaloid, has been extracted from leaves, pods, and seeds. The $LD_{50}$ orally in mice is 50-101 mg/kg, subcutaneous in dogs it is 4 mg/kg body weight.

**CONFUSED TAXA:** The bipinnate leaves of Kentucky coffee-tree may be confused with the bipinnate or even-pinnate leaves of honey locust (*Gleditsia triacanthos* L.), also in the Caesalpinaceae. Honey locust has leaflets 1 cm wide, thorns that may be branched, and flowers in spikelike racemes. The pulp of the honey locust fruit is sweet-tasting and nonpoisonous.

**SPECIES OF ANIMALS AFFECTED:** Kentucky coffee-tree is poisonous for apparently all classes of livestock and for humans.

**TREATMENT:** (11a)(b); (26)

**OF INTEREST:** A case exists of a woman who was poisoned after eating some fruit pulp of Kentucky coffee-tree, mistaking it for honey locust. Cases have been cited in which death occurred in sheep in less than 1 day after the appearance of symptoms. Seeds of this plant have been used for a coffee substitute.

The golden chain tree, *Laburnum anagyroides* Medic., a cytisine- producing legume in the Fabaceae, can cause toxicosis as well. This widely cultivated tree with long-petioled, trifoliate, alternate leaves produces hanging racemes, about 5 dm long, of golden-yellow flowers. The fruit is a legume pod containing up to 8 seeds. The seeds and flowers, eaten in large quantities, can produce poisoning characterized by excitement, gastroenteritis, dilation of pupils, incoordination, irregular pulse, convulsions, coma, and death through asphyxiation. Oral toxicity of seeds for horses is about 0.05% of the animal's weight. Treatment: (11a)(b); (26); (1); (6).

# Hedera

*H. helix*

# GENUS: *Hedera*

*Hedera helix* L. — English ivy

**FAMILY:** Araliaceae — the Ginseng Family

This family is of minor economic importance. Some noteworthy members include English Ivy (*Hedera helix* L.); *Tetrapanax papyriferus* (Hook.) C. Koch, the source of Chinese rice-paper; and *Panax quinquefolium* L., the ginseng of Oriental medicine. Because only *Hedera helix* is of major concern here, a full description of this species replaces the family description.

**PHENOLOGY:** English ivy produces umbels of flowers in summer.

**DISTRIBUTION:** *Hedera helix* is a cultivated plant grown indoors as a pot subject or outside, usually as a wall or ground cover.

**PLANT CHARACTERISTICS:** English ivy is a trailing or climbing **vine** with a diversity of leaf shapes ranging from ovate, rotund to variously 3- to 5-lobed or angled, **leaves:** firm, evergreen; **flowers:** small, greenish, produced only when the branches reach a height of more than 15 feet; **sepals:** 5, very short; **petals:** 5, fleshy; **stamens:** 5; **ovary:** 5-celled, 1 style; **fruit:** a round, 3- to 5-seeded berry.

**POISONOUS PARTS:** The black berries and leaves of English ivy are poisonous if consumed in quantity.

**SYMPTOMS:** *Hedera helix* is a purgative that produces local irritation, excessive salivation, nausea, excitement, difficulty in breathing, severe diarrhea, thirst, and coma.

**POISONOUS PRINCIPLES:** The toxic substance is hederin, a glycoside of the steroidal saponin hederagenin.

**CONFUSED TAXA:** Several varieties of *Hedera helix* have 3 leaflets per leaf and resemble poison ivy (see *Rhus radicans*). There exist numerous foliage forms; many are not stable and, with age, revert to the original type. Generally, English ivy can be differentiated from poison ivy by its dark, glossy, evergreen foliage, compared to the thinner-textured, deciduous leaves of poison ivy.

**SPECIES OF ANIMALS AFFECTED:** Both humans and livestock show the symptoms listed above.

**TREATMENT:** (11a)(b); (17); (6)

**OF INTEREST:** Other species of *Hedera*, especially the popular Algerian ivy, *H. canariensis* Willd., as well as members of the genus *Aralia* (sarsaparilla) should be viewed with suspicion. The fruits of all species of *Aralia* are poisonous when eaten raw but are infrequently cooked as jelly, which is reported edible.

# Helenium

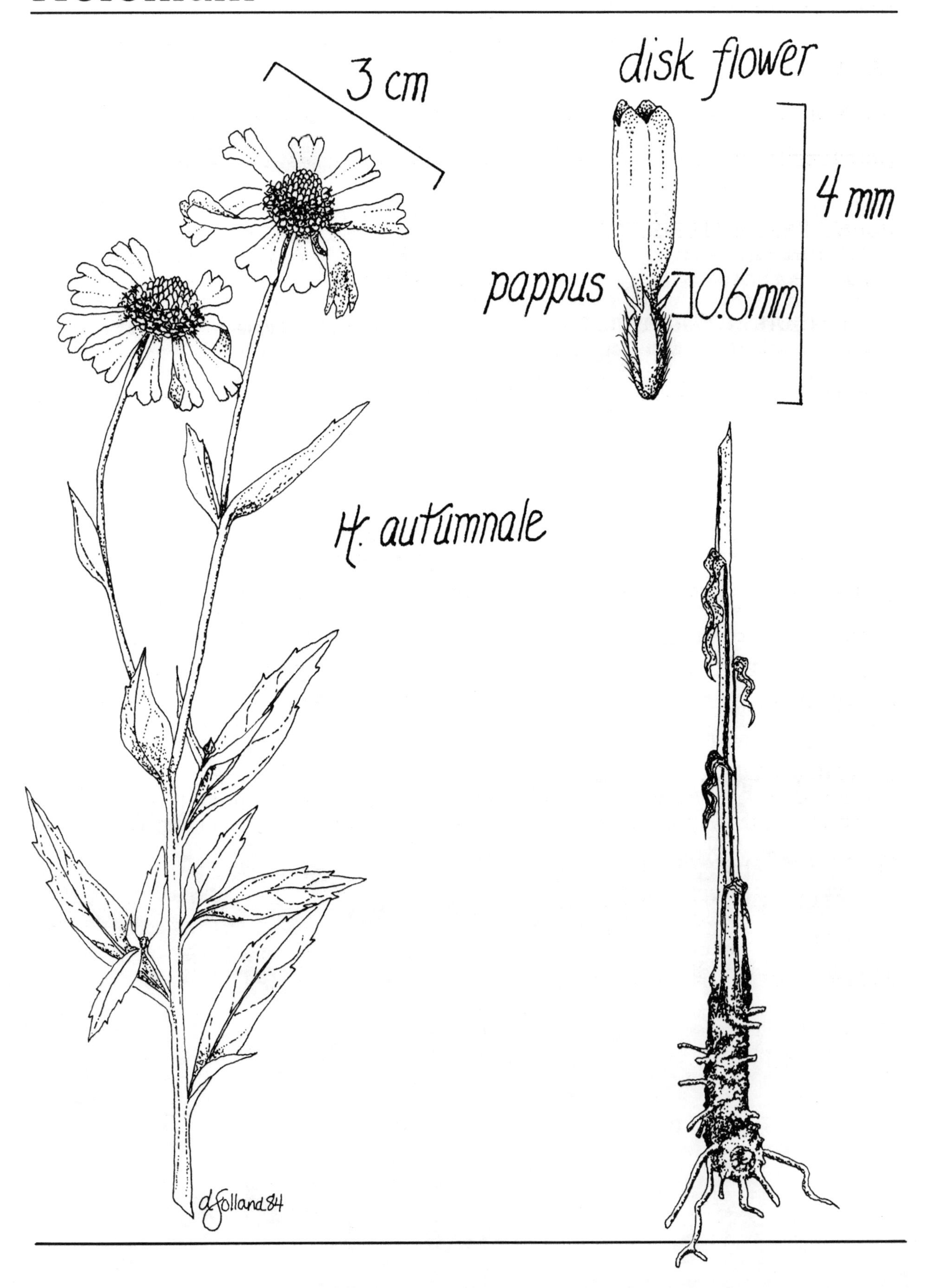

# GENUS: *Helenium*

*Helenium autumnale* L. — Sneezeweed

**FAMILY:** Compositae (Asteraceae) — the Daisy Family (see *Arctium*)

**PHENOLOGY:** *Helenium* species flower late in the growing season, from August through October.

**DISTRIBUTION:** Sneezeweed is an inhabitant of moist low ground, rich thickets, meadows, and shores.

**PLANT CHARACTERISTICS:** The fibrous-rooted perennial *H. autumnale* grows to 1.5 m tall. The **stems** bear wings originating as decurrent leaf petioles; **leaves:** numerous, lance-linear to elliptic, almost sessile; lower leaves deciduous; **heads:** several or many in a leafy corymbosely branched inflorescence, or simple; **disks:** yellow, 8-20 mm wide; **rays:** toothed or lobed, 10-20, pistillate or sometimes neutral, 1.5-2.5 cm long and 7-12 mm broad; **pappus:** keeled scales, brown, ovate, tapering into a short awn.

**POISONOUS PARTS:** All parts of the sneezeweed plant, especially flowers, are toxic. Experiments show 1% (dry) of a sheep's weight of *Helenium* will cause illness and death within 8 days.

**SYMPTOMS:** Early signs of *Helenium* poisoning are dullness and depression; weakness, tremors, and rapid respiration and pulse ensue. Nausea and vomiting may be present. Other symptoms include excessive salivation, belching, frothing, and intestinal disorders. Animals not displaying vomition will often recover. The prognosis for those vomiting is less positive.

**Postmortem: gross and histological lesions:** examination reveals gastrointestinal degeneration and liver, kidney, and lung damage; hydrothorax, ascites, congestion, and edema in the forestomachs (submucosa of the rumen and reticulum); edema of nervous tissue; and sometimes mild tubular nephrosis, as well as fatty changes in the myocardium.

**POISONOUS PRINCIPLES:** The toxin in some species is a glycoside, dugaldin, whereas in other taxa it is the sesquiterpene lactone helenatin.

**CONFUSED TAXA:** The combination of yellow ray and disk flowers, scaly-awned pappus, alternate leaves, and truncate style-branches without appendages are unique to *Helenium*. The closely related genus *Gaillardia* is similar except the style-branches have subulate appendages.

**SPECIES OF ANIMALS AFFECTED:** Sheep are apparently very sensitive to sneezeweed, but the poisonous principle will affect all livestock and humans. Sheep will eat this bitter weed when all other forage is unavailable. The toxin remains poisonous when dried; therefore, contaminated hay is also undesirable.

**TREATMENT:** (11a)(11b); (26)

**OF INTEREST:** In addition to *H. autumnale*, several less common, introduced taxa are found in Pennsylvania: *H. amarum* (Raf.) H. Rock, *H. flexuosum Raf.*, and *H. quadridentatum* Labill.

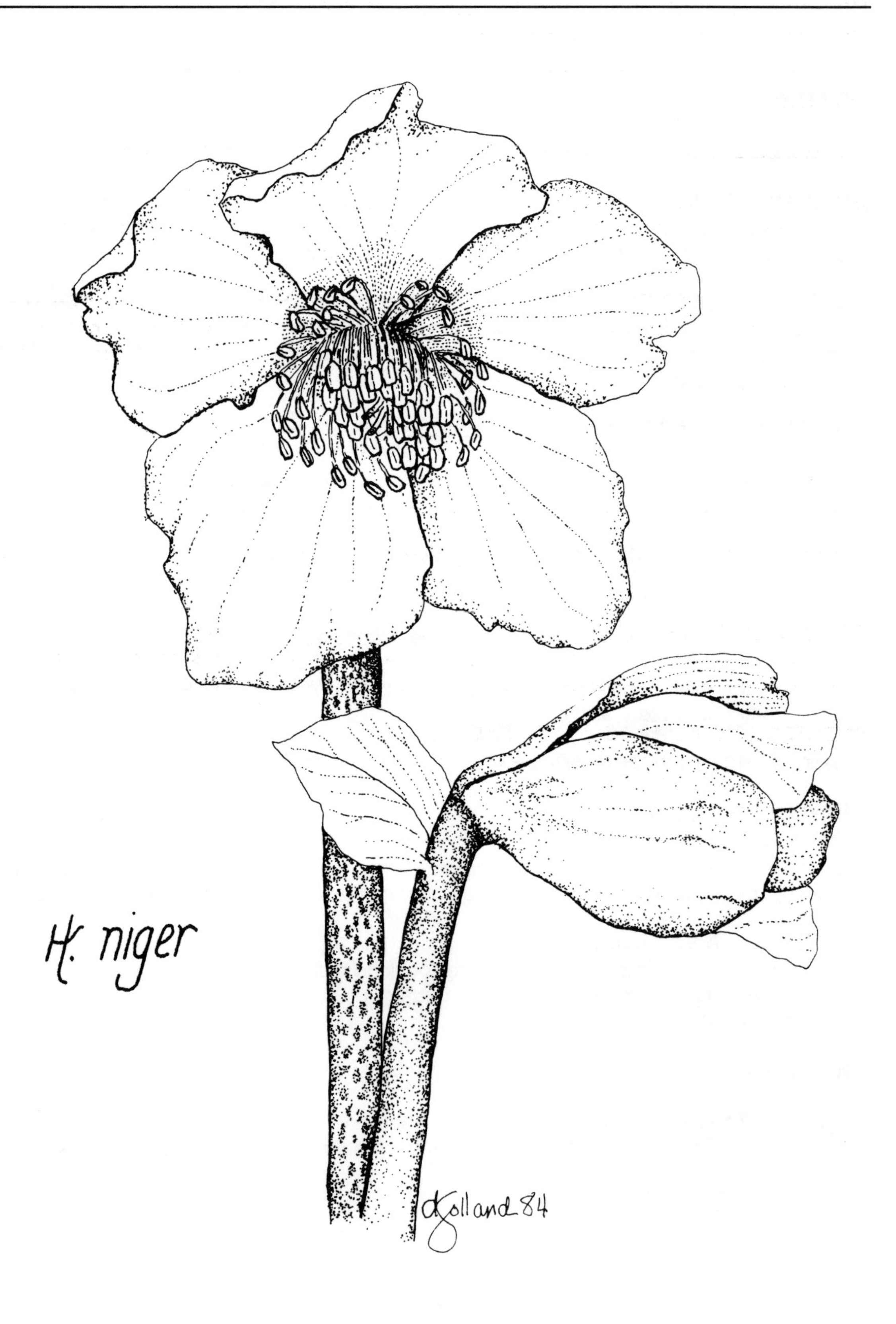
H. niger
dGolland 84

# GENUS: *Helleborus*

*Helleborus viridis* L. — Green hellebore; winter-aconite
*Helleborus niger* L. — Christmas rose

**FAMILY:** Ranunculaceae — the Buttercup Family (see *Actaea*)

**PHENOLOGY:** The hellebores are late winter to early spring flowering plants.

**DISTRIBUTION:** *Helleborus viridis* is a European species that is established in some areas. *Helleborus niger* is a durable, cold-hardy, evergreen, ornamental plant cultivated in gardens. Both thrive in partially shaded, moist situations in good soil.

**PLANT CHARACTERISTICS:** The genus *Helleborus* is recognized by showy **flowers** of white, green, or purple; **sepals:** 5, large, petaloid; **petals:** none; **stamens:** numerous, the outer 8-10 modified into staminodes; **pistils:** usually 3 or 4; **style:** erect, slender; **fruit:** a follicle; **leaves:** alternate, palmately cleft.

**POISONOUS PARTS:** The entire plant is toxic.

**SYMPTOMS:** *Helleborus* poisoning includes vomiting, diarrhea, and nervous system disturbances such as delirium, convulsions, and death due to respiratory collapse.

**POISONOUS PRINCIPLES:** Cardiac glycosides are responsible for poisonings. Hellebrin is a cardiac stimulant found in these plants.

**CONFUSED TAXA:** *Helleborus niger,* the evergreen, cold-resistant plant, produces a floral stalk but no true leafy stems. The flowers, usually borne singly on red-spotted peduncles, are white (suffused with pink). *Helleborus viridis* produces an erect stem with 2-4 drooping green flowers.

**SPECIES OF ANIMALS AFFECTED:** Both animals and humans are affected by this poisonous plant. When eaten, the hellebores are said to have a "burning taste."

**TREATMENT:** (11a)(b); (26)

# Heracleum

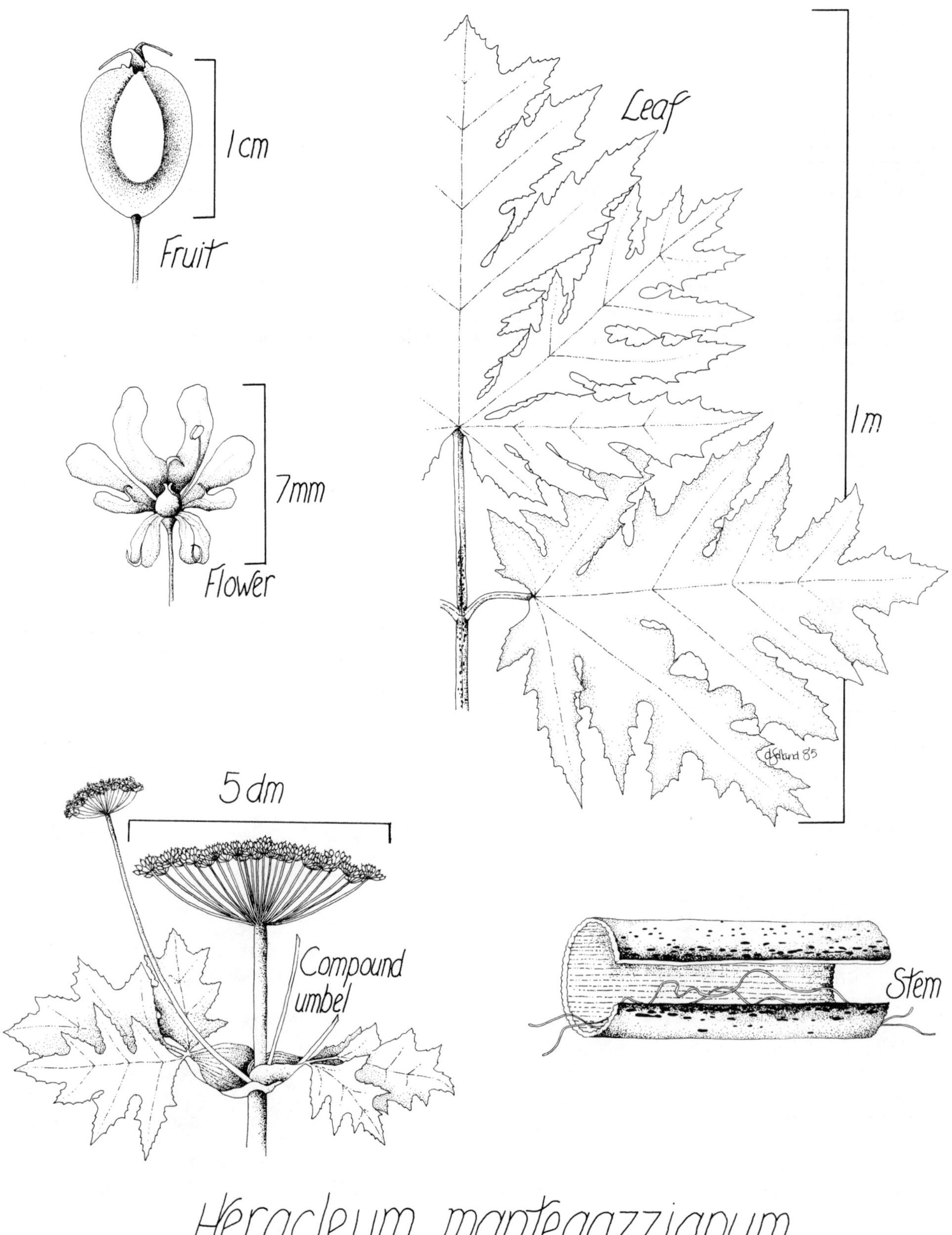


*Heracleum mantegazzianum*

# GENUS: *Heracleum*

*Heracleum lanatum* Michx. — Cow parsnip
*Heracleum mantegazzianum* Sommier & Levl. — Giant hogweed

**FAMILY:** Umbelliferae (Apiaceae) — the Umbel Family (see *Cicuta*)

**PHENOLOGY:** *Heracleum* species flower throughout June and July.

**DISTRIBUTION:** Giant hogweed is native to the Caucasus Mountains and was cultivated in the U.S. as an unusual ornamental. Cow parsnip is a native plant growing in rich moist soil and low ground.

**PLANT CHARACTERISTICS:** The genus *Heracleum* contains plants that are tall, stout perennials with large compound leaves and broad, flat-topped, compound **umbels** with deciduous involucres and many-leaved involucels; **corolla:** white, peripheral flowers of the marginal umbellets irregular with the outer, bifid corolla-lobes enlarged; **fruit:** elliptic to obovate, dorsally strongly flattened, the lateral ribs broadly winged; oil-tubes 2-4 on the commissure, extending to about half-way down the fissure; **leaves:** lower ones, once pinnate; upper ones, once ternate.

**POISONOUS PARTS:** Upon contact, the herbage and fruits are highly irritating under the conditions described below.

**SYMPTOMS:** Giant hogweed produces severe, painful, burning blisters in susceptible people, the symptoms appearing within 24 to 48 hours after contact. The sap can produce painless red blotches that later blacken and scar the skin for several years. For an adverse reaction to occur the skin, contaminated with plant juices, must be moist and subsequently exposed to sunlight (see also *Lantana* and *Hypericum*). This phenomenon, known as phytophotosensitization, occurs in animals when chemical compounds, either derived directly from plants or produced by the animal in response to plant substances, are present in peripheral circulation. *Heracleum lanatum* has also been implicated in less severe photosensitization reactions in some people.

**POISONOUS PRINCIPLES:** The glycoside furanocoumarin is responsible for the severe contact dermatitis.

**CONFUSED TAXA:** *Heracleum mantegazzianum* is an herb that grows to 4 meters with leaves sometimes over 1 meter long, leaflets: very large, deeply cut, green beneath; umbels: up to 1 meter across; petioles: blotched with purple, having large, coarse white hairs at the base; plants: coarsely hairy; flower stalks: ribbed. *Heracleum lanatum* is a smaller, less coarse (more softly pubescent) plant growing to 2 meters. Little or no purple markings are evident on the plant. *Angelica atropurpurea,* the purple-stemmed angelica, also a member of the Umbelliferae, has uniformly purple, hairless stems, and smaller, white-flowering flat-topped umbels less than 0.3 m in diameter.

**SPECIES OF ANIMALS AFFECTED:** Reports in the United States concern contact dermatitis in humans. However, photosensitization reactions can happen to livestock and pets.

**TREATMENT:** Washing sap-exposed skin with soap and water may help; where blisters appear: (4); (23); (26)

**OF INTEREST:** The recent detection of *Heracleum mantegazzianum* in Erie County provided the first record for this species in Pennsylvania. Other northern tier counties should remain alert for its presence; populations of the plant are known to occur 3 miles north of the Pennsylvania-New York border. It grows along roadside ditches and in moist waste areas. It has become naturalized in at least two dozen counties in central and western New York. *Heracleum lanatum* is reported to have medicinal properties and was once used for treating epilepsy.

# Hydrangea

# GENUS: *Hydrangea*

*Hydrangea arborescens* L. — Hills-of-snow; sevenbark
*Hydrangea macrophylla* (Thunb.) Ser. — Hydrangea
*Hydrangea paniculata* Siebold — PeeGee hydrangea

---

**FAMILY:** Saxifragaceae — the Saxifrage Family

The only member of this large and diverse family with known poisonous properties is *Hydrangea.* The characteristics of *Hydrangea,* provided below, replace the family description.

**PHENOLOGY:** The introduced common garden hydrangea, *H. macrophylla,* flowers in mid-summer. The native tree hydrangea or hills-of-snow *(H. arborescens)* is found blooming June and July, while the introduced PeeGee hydrangea *(H. paniculata)* flowers later in the season, during August and September.

**DISTRIBUTION:** The garden hydrangea is a well known, "old fashioned" ornamental cultivated for display in outdoor plantings, pots, or tubs. The hills-of-snow is a garden cultivar of the native *H. arborescens* that is found on dry or moist rocky woods and hillsides. The PeeGee hydrangea, *H. paniculata,* is a native of eastern Asia, also used in landscaping.

**PLANT CHARACTERISTICS:** The hydrangeas mentioned above have **leaves:** deciduous, opposite, petioled, toothed; **flowers:** in flat-topped or globular terminal cymes, often with the outer flowers sterile and much enlarged relative to the inner ones, ranging in color from white to pink, lavender, or blue; **flowers:** 5-merous; **sepals:** showy; **ovary:** inferior, or nearly so; **fruits:** dry capsules containing many small seeds.

**POISONOUS PARTS:** The leaves and buds contain the poisonous constituents.

**SYMPTOMS:** Under certain conditions the toxins produce gastrointestinal upset, nausea, diarrhea (bloody), and vomiting.

**POISONOUS PRINCIPLES:** Research indicates that hydrangea sometimes contains a cyanogenic glycoside, hydrangin. Other constituents include saponin, resins, fixed and volatile oils, and starch.

**CONFUSED TAXA:** The hydrangeas are popular cultivated plants not easily confused with other taxa.

**SPECIES OF ANIMALS AFFECTED:** Both livestock and human cases have been reported. Sickness was painful but recovery occurred.

**TREATMENT:** (11a)(b); treat for cyanide poisoning.

**OF INTEREST:** Analysis of poisoned victims does not always show symptoms compatible with cyanide poisoning. Roots of *H. arborescens* were used by American pioneers in the treatment of dyspepsia.

# Hypericum

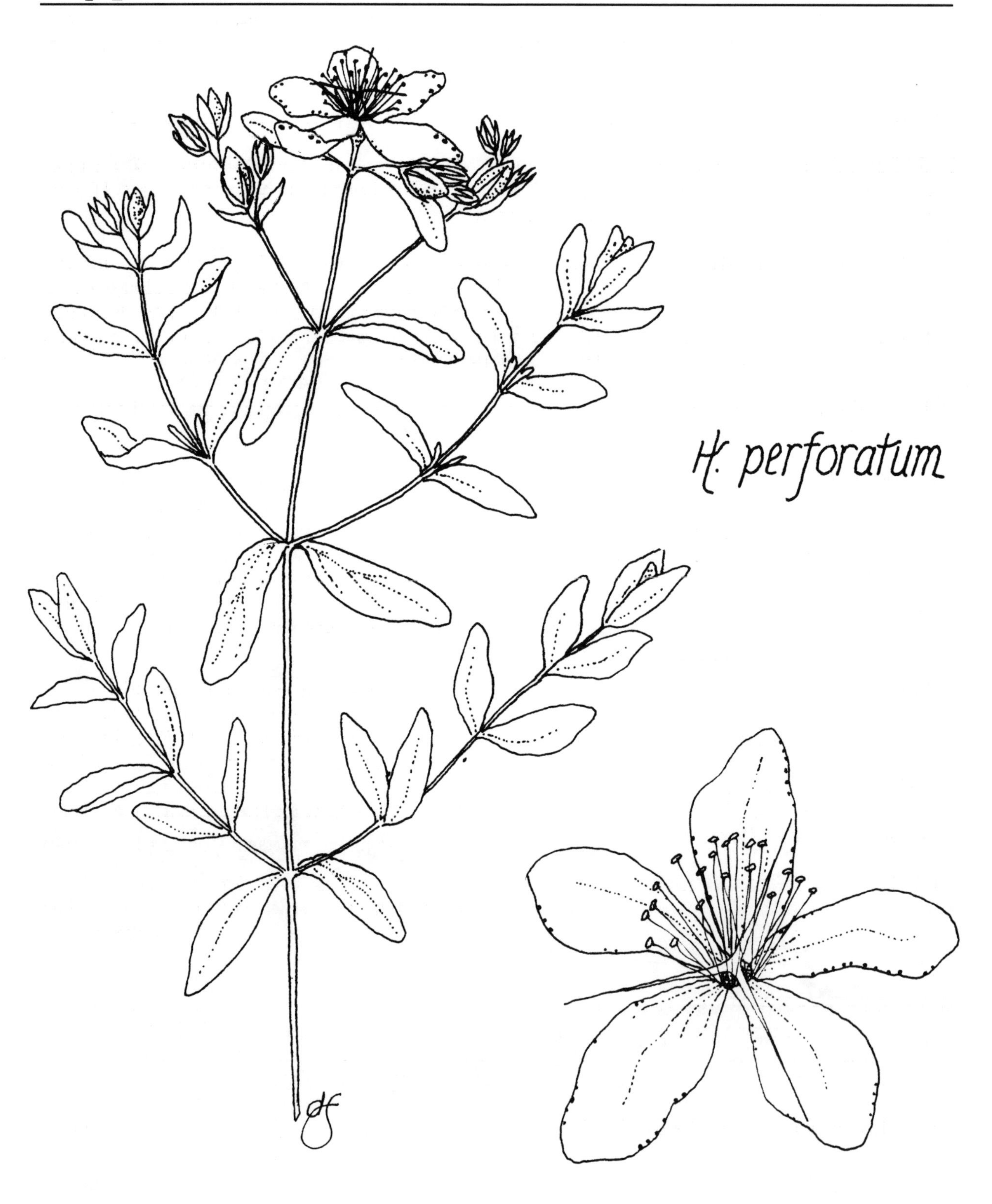


**GENUS:** *Hypericum*
*Hypericum perforatum* L. — St. John's wort

**FAMILY:** Hypericaceae — the St. John's wort Family

Native to temperate and tropical regions, this family of few genera has **leaves:** opposite or whorled, simple, entire, usually translucent-dotted or black-dotted; **flowers:** usually yellow, regular, bisexual; **petals:** separate; **stamens:** numerous, often united into clusters; **styles:** separate; **fruit:** a capsule.

**PHENOLOGY:** *Hypericum perforatum* flowers over an extended period, June through September.

**DISTRIBUTION:** The genus *Hypericum* provides several ornamentals for borders, rock gardens, ground-covers, or landscape shrubbery. *H. perforatum* has become an abundant weed of fields, meadows, roadsides, pastures, and waste places.

**PLANT CHARACTERISTICS:** *Hypericum perforatum* is a perennial, 4-8 dm, with many upright, leafy **branches** that are sharply ridged below the base of each leaf; **leaves:** linear-oblong, 2-4 cm on the main stem, half as long on the branches; **inflorescence:** cymose, flat topped, terminal; **flowers:** numerous; **sepals:** 4-6 mm, with few or no black dots; **petals:** 8-10 mm, black-dotted along the margin; **stamens:** in 3 clusters; **styles:** 3; **seeds:** 1-1.3 mm.

**POISONOUS PARTS:** All parts of the plant that bear the black dots, including petals and herbage, are poisonous.

**SYMPTOMS:** The black glands contain a toxin that is a primary phytophotosensitizer. These compounds are absorbed through the digestive system without alteration. In the circulatory system of mammals the chemicals damage the liver. In the presence of sunlight, skin develops dermatitis. Animals with normally dark pigmented skin are less likely to develop skin lesions. *Hypericum perforatum* produces intense dermal itching associated with aberrant behavior. Animals may experience convulsions prior to death. Often mucosa and eye epithelium are highly irritated. Blindness, and starvation may precede death. Additional symptoms include elevated temperature, increased respiration and heart rate, and diarrhea.

**Postmortem: gross lesions:** dermatitis, conjunctivitis; skin lesions in cattle on teats, udder, and escutcheon; in sheep on head, ears, lips, eyelids, and coronet; skin changes proceed from reddening to edema, fluid weeps from the skin under necrotic tissue, and sloughs. In cattle wounds heal slowly (approximately 2 weeks) and produce hairless scars. Death may result from infection and gangrene. Pyridine extracts of mouth, nasal and conjunctiva mucosa, and digestive tract produce a light-red fluorescence under Wood's UV light; **histological lesions:** skin lesions including hyperemia, edema, necrosis, and ulceration.

**POISONOUS PRINCIPLES:** The toxic chemical probably is hypericin, a derivative of dianthrone. Previously, two substances, a volatile oil and hypericum red, were implicated. It has recently been suggested that a pigment, probably a mixture of numerous polyhydroxy derivatives of helianthrone, are responsible for primary photosensitization.

**CONFUSED TAXA:** More than a dozen species of *Hypericum* occur as native or naturalized plants in Pennsylvania; several additional species are cultivated in gardens. Although only *H. perforatum* is reported in the literature as poisonous, numerous other species produce black, glandular dots, which may prove to contain hypericin or related phytotoxins.

**SPECIES OF ANIMALS AFFECTED:** Humans, cattle, goats, sheep, and horses are reported to be affected.

**TREATMENT:** (11a)(b); (26); avoid direct sunlight after ingestion.

**OF INTEREST:** The active compound hypericin has a tonic and tranquilizing action on humans in very small quantities. Hay contaminated with dried St. John's wort is toxic since hypericin is stable upon drying and resistant to destruction by heat.

*I. opaca*

# GENUS: *Ilex*

*Ilex* spp. — Holly

**FAMILY:** Aquifoliaceae — the Holly Family

The holly family is represented in Pennsylvania by two genera: common holly, *Ilex* and mountain-holly with a single species, *Nemopanthus mucronatus* (L.) Trel. The genus *Ilex* is of concern due to its mildly poisonous berries. The family consists of **trees** or **shrubs; leaves:** simple, alternate; **stipules:** minute, caducous; **flowers:** small, 4-merous, often unisexual. Many species are dioecious, containing plants that bear only male or female flowers.

**PHENOLOGY:** Hollies generally flower in May or June. The berries of most holly plants begin ripening in late autumn.

**DISTRIBUTION:** *Ilex* is encountered in Pennsylvania as ornamentals or as native or escaped plants. The evergreen, ornamental hollies are found in landscaping around the home where the soil is rich, acidic, and well drained. Many of the native species prefer wet woods or swamps.

**PLANT CHARACTERISTICS:** Most species of *Ilex* have **flowers:** axillary, small, greenish white, unisexual, male and female borne on separate plants; **pistillate flowers:** bear stamens with reduced anthers; **staminate flowers:** produce as many stamens as the number of petals, often with a vestigial pistil; **calyx:** 4-6 lobed; **petals:** 4-8, slightly fused at base; **fruit:** a red or black berry.

**POISONOUS PARTS:** The berries of *Ilex* are mildly poisonous.

**SYMPTOMS:** Holly berries produce gastrointestinal disturbances (vomiting, diarrhea) and stupor when consumed in large amounts.

**POISONOUS PRINCIPLES:** The toxin is unknown.

**CONFUSED TAXA:** Holly and holly berries are easily recognized by most people because of their popularity.

**SPECIES OF ANIMALS AFFECTED:** Small children are most likely to be poisoned from ingestion of holly berries.

**TREATMENT:** (11a)(b); (26)

**OF INTEREST:** The steeped berries of American holly, *Ilex opaca,* have been used by Amerindians as a cardiac stimulant. The dried leaves have been used by colonists and immigrants as a substitute for tea, especially during the American Civil War. Care should be taken during the Christmas season to keep children away from *Ilex* berries.

# GENUS: *Ipomoea*

*Ipomoea* spp. — Morning-glory

**FAMILY:** Convolvulaceae — the Morning-glory Family

This moderate-sized family contains **twining herbs,** often with milky juice; **leaves:** alternate, in our range simple and lobed; **flowers:** large, brightly colored, regular, and bisexual; **calyx:** 5-parted; **corolla:** funnelform, pleated; **buds:** frequently twisted; **stamens:** 5; **ovary:** superior.

**PHENOLOGY:** Generally all species flower July throughout September.

**DISTRIBUTION:** Some of the *Ipomoea* species are cultivated; some such as *I. purpurea* (L.) Roth have escaped and are found in uncultivated situations. Still others, such as *I. pandurata* (L.) G.F.W. Meyer, are native plants. Many occur as weeds of fields, roadsides, thickets, and waste places.

**PLANT CHARACTERISTICS:** Morning-glories have the following characteristics; **sepals:** 5, imbricate, often unequal; **stamens** and **style** not exsert from **corolla; leaves:** cordate or lobed.

**POISONOUS PARTS:** The leaves and stems are toxic, while the seeds in some species are hallucinogenic.

**SYMPTOMS:** Ingestion of vegetation causes purgation and gastrointestinal distress accompanied by explosive diarrhea, frequent urination, and depressed reflexes. Prolonged consumption results in anorexia, wasting-away, depression, dyspnea, coma, and in severe cases, death. Consumption of seeds causes nausea, psychotic reactions, and hallucination.

**POISONOUS PRINCIPLE:** The toxins in *Ipomoea* foliage are unknown. The hallucinogenic principle in seeds of *I. tricolor* Cav. is D-lysergic acid amide (ergine) and possibly other ergot alkaloids. The seeds are estimated to contain 3 mg of alkaloid per gram. This compound is similar to LSD-25, a hallucinogen.

**CONFUSED TAXA:** All native and naturalized morning-glories should be considered toxic. The seeds of the commonly cultivated *I. tricolor* and cultivars ('Blue Star', 'Flying Saucers', 'Heavenly Blue', 'Pearly Gates', 'Summer Skies', 'Wedding Bells', etc.) are dangerous. The genus *Convolvulus* (bindweed) differs from *Ipomoea* in that the former has 2 stigmas, whereas the latter has one. Consumption of bindweed foliage is also reported to cause gastric distress.

**SPECIES OF ANIMALS AFFECTED:** Livestock have developed the symptoms described above after consuming leaves and stems. Especially susceptible are hogs, sheep, cattle, and goats. Humans have been poisoned from an overdose of hallucinogenic seeds.

**TREATMENT:** (11a)(b); (26)

**OF INTEREST:** As a hallucinogen, morning-glory seeds are estimated to be about 1/10 as potent as LSD.

It should be noted that many seed distributors apply fungicides to seeds prior to packaging. Consumption of treated seeds can cause additional sickness, including severe vomiting, diarrhea, or physiological problems.

In Mexico the Aztecs used seeds of *Ipomoea* and the related genus *Rivea* as a hallucinogen in religious ceremonies and in medicine.

# GENUS: *Iris*

*Iris* spp. — Iris; flag

---

**FAMILY:** Iridaceae — the Iris Family

In addition to the well known cultivated and native or naturalized species of *Iris,* other members of the family found in our range include the blackberry-lily, *Belamcanda chinensis* (L.) DC.; gladeolus, *Gladeolus* spp.; and blue-eyed grass, *Sisyrinchium* spp. Owing to the familiarity of this group, especially the irises, no family description is given.

**PHENOLOGY:** The flowering time of irises varies, with early-flowering species producing showy blossoms in April and others blooming as late as July. Generally, most irises in our range flower in May.

**DISTRIBUTION:** Irises thrive in habitats ranging from sandy, open woods to swamps. Many types are garden cultivars, grown around the home for ornamental purposes.

**PLANT CHARACTERISTICS:** The petals and sepals are not readily distinguishable; **outer tepals:** spreading or reflexed; **inner tepals:** erect or arching; **stamens:** inserted at base of outer 3 tepals; **ovary:** 3-6 angled; **style:** divided distally into 3 petaloid branches arching over the stamens, each 2-lobed at the tip; **perennial herbs** with linear **leaves** growing from a horizontal rhizome.

**POISONOUS PARTS:** The roots, and to a lesser extent the leaves, are poisonous upon ingestion in quantity. The roots may produce dermatitis in sensitive individuals.

**SYMPTOMS:** Ingestion of iris leads to gastroenteritis, purgation, and dyspnea. Contact dermatitis and irritation may result.

**POISONOUS PRINCIPLES:** The toxins are largely unknown. The irritant principle may be irone, a glycoside.

**CONFUSED TAXA:** The flowers of iris are unmistakable. The leaves may be confused with cat-tail (*Typha* spp.), sweet flag (*Acorus* spp.), or some larger sedges (*Carex* spp.).

**SPECIES OF ANIMALS AFFECTED:** Livestock and humans have been poisoned after eating large quantities of iris.

**TREATMENT:** (11a)(b); (26) for ingestion; (23) for dermatitis

**OF INTEREST:** Toxic reports are rare, probably due to the large quantity of material needed for poisoning. Iris root has been used medicinally as a purgative. Chemical constituents include iridin (an oleoresin), isophthalic acid (an unknown camphoraceus substance), gum, tannin, sugars, and oils. The seeds of iris have been used as a coffee bean substitute, a practice not advised. The pulped raw rhizome of *I. missouriensis* was used to relieve tooth ache.

*Gladeolus* spp. have been listed in older literature as poisonous, but this could not be substantiated in recent references to poisonous plants.

# Kalmia

**GENUS:** *Kalmia*

*Kalmia latifolia* L. — Mountain laurel

**FAMILY:** Ericaceae — the Heath Family

Widely distributed on acid soils, members of this family are found mostly in the northern temperate region. Characteristics include:. **flowers:** regular, bisexual, and perfect; **petals:** usually united; **leaves:** alternate, opposite, or whorled on the stems; **stamens:** as many or twice as many as the petals; **calyx:** 4-7 lobed, often 5; **corolla:** often urceolate; **anthers:** often appendaged, frequently opening by a terminal pore; **pistil:** 1; **carpels:** 5; **style:** 1; **fruit:** a capsule.

**PHENOLOGY:** Mountain laurel flowers May through July.

**DISTRIBUTION:** Woodlands on rocky or sandy acidic soil.

**PLANT CHARACTERISTICS:** Shrubs or small trees, 2 to 10 m high; **petioles:** 1-2 cm; **leaves:** evergreen, alternate, glabrous, 5-10 cm long, margin entire, dark green above, bright green below; **flowers:** terminal; **corolla:** white to rose with purple markings; **anthers:** held in chambers on the corolla tube until pollination; **fruit:** a dry, 5-celled septicidal capsule.

**POISONOUS PARTS:** Flowers, twigs, pollen grains, and green plant parts cause toxicity. Percentages of *Kalmia* (relative to animal's body weight) needed to produce symptoms, but not death, are: 0.15% (sheep), 0.2-0.4% (cattle and goats), and 1.3% (deer).

**SYMPTOMS:** In order of appearance, symptoms are: anorexia; repeated swallowing or eructation and swallowing of cud without mastication; profuse salivation; watering of the mouth, eyes, and nose; loss of energy; slow pulse; low blood pressure; incoordination; dullness; depression; vomiting; and frequent defecation. As poisoning progresses, animals become weak and prostrate. Difficulty in breathing is common and there is no pupillary reflex; death is preceded by coma. Symptoms are similar for all classes of livestock; the time for the appearance of symptoms averages 6 hours. **Postmortem:** nonspecific gastrointestinal irritation and hemorrhage.

**POISONOUS PRINCIPLES:** Andromedotoxin and arbutin are responsible for toxicity. Andromedotoxin a resinoid, causes vomiting by directly stimulating the vomition center; its structure is not fully known. Arbutin is a glucoside by hydroquinone.

**CONFUSED TAXA:** Three species of *Kalmia* occur in Pennsylvania: *K. angustifolia* L. with lateral flower clusters and *K. polifolia* Wang and *K. latifolia* L. with terminal flower clusters. *Kalmia polifolia* has opposite leaf arrangement (as does *K. latifolia,* which can also have ternate leaves), whereas *K. latifolia* has alternate leaves.

**SPECIES OF ANIMALS AFFECTED:** Apparenty all species of animals can be poisoned by mountain laurel. Sheep are the most susceptible of animals, cattle are next. Monkeys, angora goats, and humans have been poisoned by mountain laurel.

**TREATMENT:** (11a)(b); (1); (5); (12)

**OF INTEREST:** The Delaware Indians used laurel for suicide. Experiments indicate that fowl can eat relatively large quantities of mountain laurel without developing symptoms. When fed to cats, the meat was toxic. Humans have been poisoned by andromedotoxin in pollen after eating honey suspected to have been made from members of the Ericaceae.

The toxin arbutin is used commercially as a stabilizer for color photographic images and in veterinary science as a diuretic and urinary anti-infective. Many members of this family contain these or similar toxins. Evergreen plants are more commonly the cause of poisoning than the deciduous species. Plants in this family that occur in Pennsylvania and cause sickness or death similar to that described above include *Ledum groenlandicum* Oed., Labrador tea; *Pieris japonica* (Thunb.) D. Don, Lily-of-the-valley bush (commonly cultivated); and possibly *Menziesia pilosa* (Michx.) Juss.

# Lantana

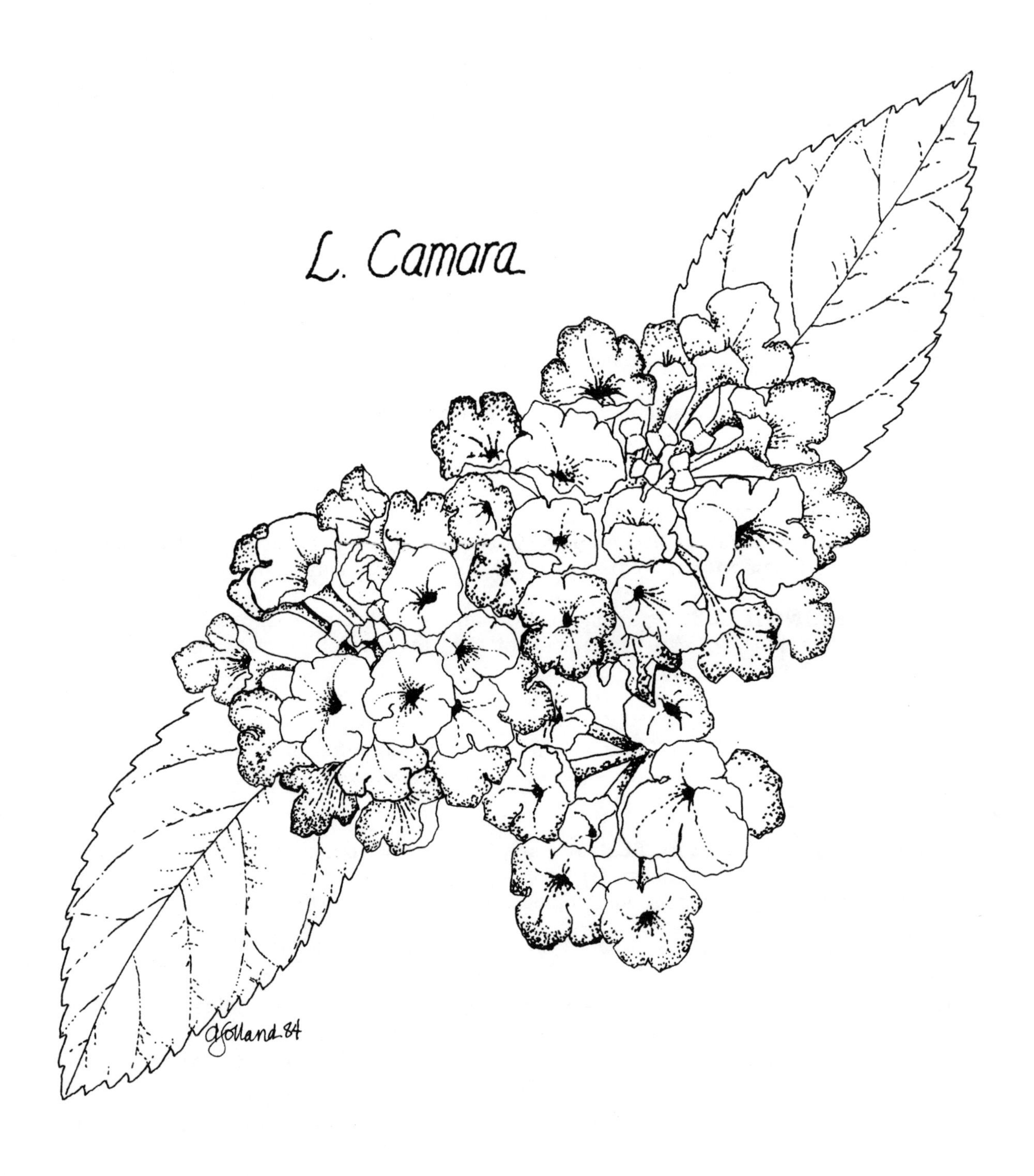

# GENUS: *Lantana*

*Lantana Camara* L. — Lantana; yellow sage; red sage

**FAMILY:** Verbenaceae — the Vervain Family

This family is widespread in the tropics but sparingly represented in cool regions. None of the twelve or more species of *Verbena* occurring in Pennsylvania are toxic. The plant in this family that can be a problem is *Lantana Camara,* a common florist's subject, propagated in greenhouses and hanging-baskets. In Florida it is one of the most common causes of poisonings. Because no other members of the Verbenaceae are toxic, a family description is replaced by the plant characteristics listed below.

**PHENOLOGY:** *Lantana* is propagated by softwood cuttings and seeds. In our area it is a hot-house plant that is flowering when purchased, usually during spring and summer months.

**DISTRIBUTION:** *Lantana Camara* is found in homes, shopping malls, and greenhouses. It is most frequently sold in hanging-baskets.

**PLANT CHARACTERISTICS:** As a potted plant lantana becomes bushy and produces abundant flowers; **stems:** square; **leaves:** opposite, ovate, crenate-dentate, to 25 cm long, rough above, aromatic when crushed; **inflorescences:** flat-topped heads to 5 cm across, orange-yellow or orange changing to red, or white; **peduncles:** axillary, longer than the leaves; **flowers:** tubular, 4-parted, small; **fruits:** black (greenish when immature), fleshy, one-seeded drupes, 8 mm in diameter.

**POISONOUS PARTS:** The green, unripened fruit is very dangerous. Leaves of lantana also have yielded toxic principles upon extraction. Feeding studies indicate that lantana is quite poisonous. About 1% (green-weight basis) of body weight is sufficient for bovine toxic reactions. Fresh lantana fed to sheep produced acute symptoms and death within 5 days at about 2% of the animal's weight.

**SYMPTOMS:** Toxicosis produces gastrointestinal distress, vomiting, bloody-watery diarrhea, muscular weakness, ataxia, visual disturbances, lethargy, circulatory failure, and death. In acute cases lantana toxicity resembles atropine poisoning. The degree of poisoning depends on the amount of plant consumed and the degree of exposure to sunlight. Lantana contains toxins that cause organisms to react when exposed to the sun (photosensitization, see also *Heracleum* and *Hypericum*). **Postmortem** studies reveal degenerative changes in the liver and lesions of gastroenteritis. Edema and hemorrhages in some organs can occur in chronic cases.

**POISONOUS PRINCIPLES:** The alkaloid lantanin and a triterpene derivative, lantadene A, are implicated in poisonings.

**CONFUSED TAXA:** No other hanging-basket or potted plant from the greenhouse has the characteristics described.

**SPECIES OF ANIMALS AFFECTED:** Children have died after consumption of unripened berries. Beef and dairy cows are reported to succumb to browsing of lantana.

**TREATMENT:** (11a)(b); (26)

# Lathyrus

# GENUS: *Lathyrus*

*Lathyrus* spp. — Vetchling; wild pea; flat pea

---

**FAMILY:** Fabaceae (Leguminosae) — the Bean Family (see *Crotalaria*)

**PHENOLOGY:** The vetchlings flower in summer, generally June through September. A few species flower earlier.

**DISTRIBUTION:** The genus *Lathyrus* is represented by at least a half dozen species in Pennsylvania. The native taxa generally are found in a diversity of habitats: *L. ochroleucus* L., dry or moist soil, slopes and rocky banks; *L. japonicus* Willd., gravelly shores; *L. palustris* L., shores, damp thickets, meadows; *L. venosus* Muhl., rich woods, thickets, and banks of streams. The introduced taxa (*L. aphaca* L., *L. tuberosa,* and *L. latifolius* L.) either escape from cultivation to roadsides, thickets, and wasteplaces or are grown for ornamental value.

**PLANT CHARACTERISTICS:** Our species of *Lathyrus* are characterized by plants that vine; **stems:** winged or angular; **leaves:** alternate, even-pinnate, terminating in **tendrils; stamens:** diadelphous; **pods:** flat, dehiscent legumes.

**POISONOUS PARTS:** Of primary concern is the pealike seed of some species. The foliage will also produce symptoms.

**SYMPTOMS:** Even moderate amounts of *Lathyrus* seeds in the diet do not produce poisonings. The development of lathyrism is apparent after consumption of large quantities or an exclusive diet of seeds. Lathyrism is well documented in human history when war, poverty, or drought have altered the diet of the people in a region. Human symptoms include paralysis (with loss of bladder or bowel control); slow, weak pulse; muscle tremors; a posture of feet turned-in, toes down; sensory disturbances; convulsions; and death.

Horses probably are the animals most sensitive to the toxic principles. They display symptoms similar to those cited above and also hind leg paralysis, dyspnea, and roaring. In toxicity experiments in rats, *L. latifolius* produced nervous symptoms of hyperexcitability, convulsions, and death.

**POISONOUS PRINCIPLES:** Numerous compounds have been extracted from various species of *Lathyrus*. One compound, L -alpha, gamma-diaminobutyric acid, isolated from *L. latifolius,* produces the symptoms described in the rat experiments noted above.

**CONFUSED TAXA:** The vetchlings *(Lathyrus),* which resemble the vetches *(Vicia),* differ in having wing petals separate from keel petals, and a flattened style, bearded on the inner face.

**SPECIES OF ANIMALS AFFECTED:** Both humans and livestock have been poisoned by *Lathyrus*.

**TREATMENT:** (11a)(b); (26); a change of diet, thereby removing the toxic principles, can alter the progress of the poisoning.

# Ligustrum

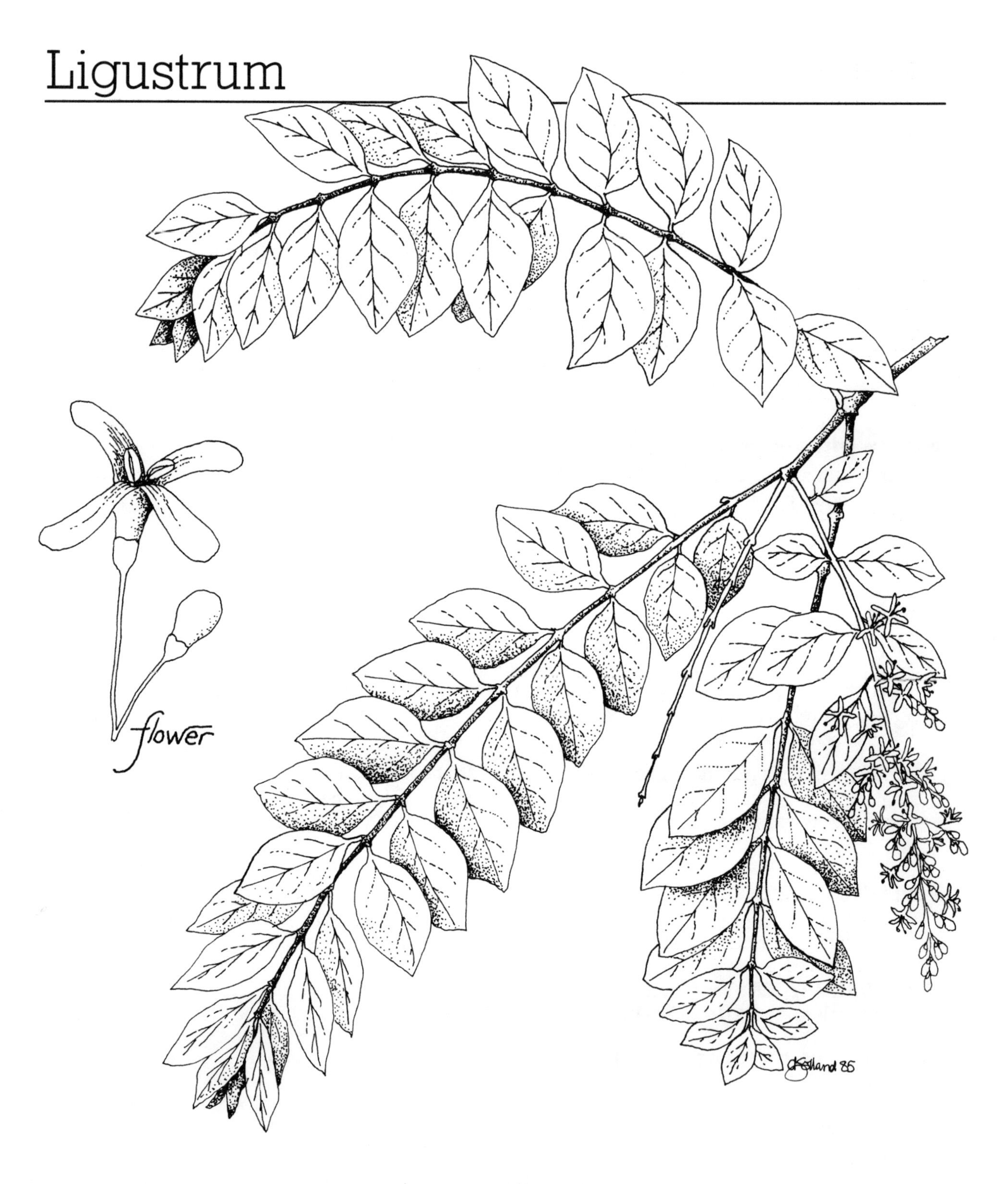


*L. vulgare*

# GENUS: *Ligustrum*

*Ligustrum vulgare* — Privet

**FAMILY:** Oleaceae — the Olive Family

In Pennsylvania this small family is represented by both native and introduced species and contains such non-toxic plants as our familiar ash trees *(Fraxinus)* and the cultivated plants forsythia *(Forsythia)* and lilac *(Syringa)*. Economically, the family contributes plant material for ornamental and shade use, and in warmer climates provides the Mediterranean olive *(Olea)* for its edible fruit. The privet hedges *(Ligustrum)* produce poisonous berries and are included in this treatment. Family characteristics are **leaves:** primarily opposite; **calyx:** commonly 4-lobed; **corolla:** 4-lobed; **stamens:** 2; **ovary:** superior, 2-celled; **fruit:** a berry, drupe, capsule, or samara.

**PHENOLOGY:** *Ligustrum* produces intensely strong-scented flowers in June and July.

**DISTRIBUTION:** Four species of privet, all introductions from Europe or Asia, are cultivated in Pennsylvania. They escape to thickets, open woods, roadsides, and borders of woodlands.

**PLANT CHARACTERISTICS:** These woody deciduous **shrubs,** commonly cultivated for hedges, have handsome foliage and a profusion of white flowers, **leaves:** opposite, simple, entire, often thick and lustrous-green, oblong or ovate, 2.5-6 cm long; **flowers:** in pyramidal panicles terminating branches and branchlets; **calyx:** short-tubular, 4-toothed; **corolla:** 4-lobed; **stamens:** 2, inserted on the corolla tube; **berry:** black when mature, 1-2 seeded, hard, becoming dry and papery.

**POISONOUS PARTS:** The vegetation and berries are poisonous.

**SYMPTOMS:** The foliage and fruit produce severe gastroenteritis, pain, vomiting, and death.

**POISONOUS PRINCIPLES:** In *Ligustrum* the toxin is believed to be an unknown glycoside.

**CONFUSED TAXA:** This common hedge is familiar to many. No other cultivated or native shrub appears like privet, described above.

**SPECIES OF ANIMALS AFFECTED:** Although poisoning is rare, humans (children), as well as horses, sheep, and cattle, have suffered from consumption of *Ligustrum*.

**TREATMENT:** (11a)(b); (26)

L. inflata
dHolland 84

# GENUS: *Lobelia*

*Lobelia cardinalis* L. — Cardinal-flower
*Lobelia inflata* L. — Indian tobacco
*Lobelia spicata* Lam. — Lobelia

---

**FAMILY:** Lobeliaceae — the Lobelia Family

Although this family is mainly tropical, several members occur in Pennsylvania. The plants contain acrid, milky sap; **leaves:** alternate, simple, entire, toothed or pinnately parted; **calyx:** 5-lobed; **corolla:** irregular, 2-lipped, 5-lobed, the tube split nearly to the base on one side; **stamens:** 5; **anthers:** united into a tube around the style; **ovary:** inferior, 2- to 5-celled; **fruit:** a capsule.

**PHENOLOGY:** Lobelias generally flower July through September.

**DISTRIBUTION:** Lobelias are found in wet soil, along streams, ponds, shores, and in swamps. They are also cultivated for garden use.

**PLANT CHARACTERISTICS:** In *Lobelia* the **corolla** is characteristically split to the base on the upper side; bilabiate, having 2 lobes above and below, the upper lobes erect, the lower lobes usually spread; **stamens:** protrude through the split in the corolla; the 2 lower stamens bearded at the tip; **inflorescence:** a terminal bracteate raceme, flowers alternately inserted; **leaves:** decurrent.

**POISONOUS PARTS:** All parts of *Lobelia* are poisonous. *Lobelia* is toxic to animals at 0.5% of body weight.

**SYMPTOMS:** Toxicosis develops within 3 days. In livestock, symptoms are sluggishness, salivation, diarrhea, anorexia, ulceration around the mouth, nasal discharges, and eventually coma. Also, lesions of hemorrhage and mild gastroenteritis may be present. In humans, symptoms include vomiting, sweating, pain, paralysis, depressed temperature, rapid but weak pulse, collapse, coma, and death.

**POISONOUS PRINCIPLE:** Toxins are pyridine alkaloids, especially lobeline.

**CONFUSED TAXA:** *Lobelia cardinalis,* a tall (1 m) perennial of stream banks, produces red flowers in late summer. *Lobelia inflata* is a tall (to 1 m in moist ground, much less in dry habitats) annual, which produces pale blue to white flowers July through October. The hypanthium (calyx-tube) is much inflated in fruit. *Lobelia spicata* is a perennial resembling *L. inflata* but not developing a bladderlike hypanthium. At least a dozen species and varieties of *Lobelia* are elements of Pennsylvania's flora. All should be considered poisonous, with perhaps *L. inflata* the most toxic. Cultivated taxa include *L. erinus* L. (edging lobelia), *L.* x *Gerardii* Chab. & Gouj. ex Sauv., *L. siphilitica* L., *L.* x *speciosa* Sweet, and *L. tenuior* R. Br.

**SPECIES OF ANIMALS AFFECTED:** Humans and livestock are susceptible.

**TREATMENT:** (1 1a)(b); (26); (6) (5- at the rate of 2 mg IM as needed)

**OF INTEREST:** *Lobelia inflata* is the source of the alkaloid lobeline used medicinally as a respiratory stimulant and in veterinary science as a respiratory stimulant and ruminatonic.

# Lolium

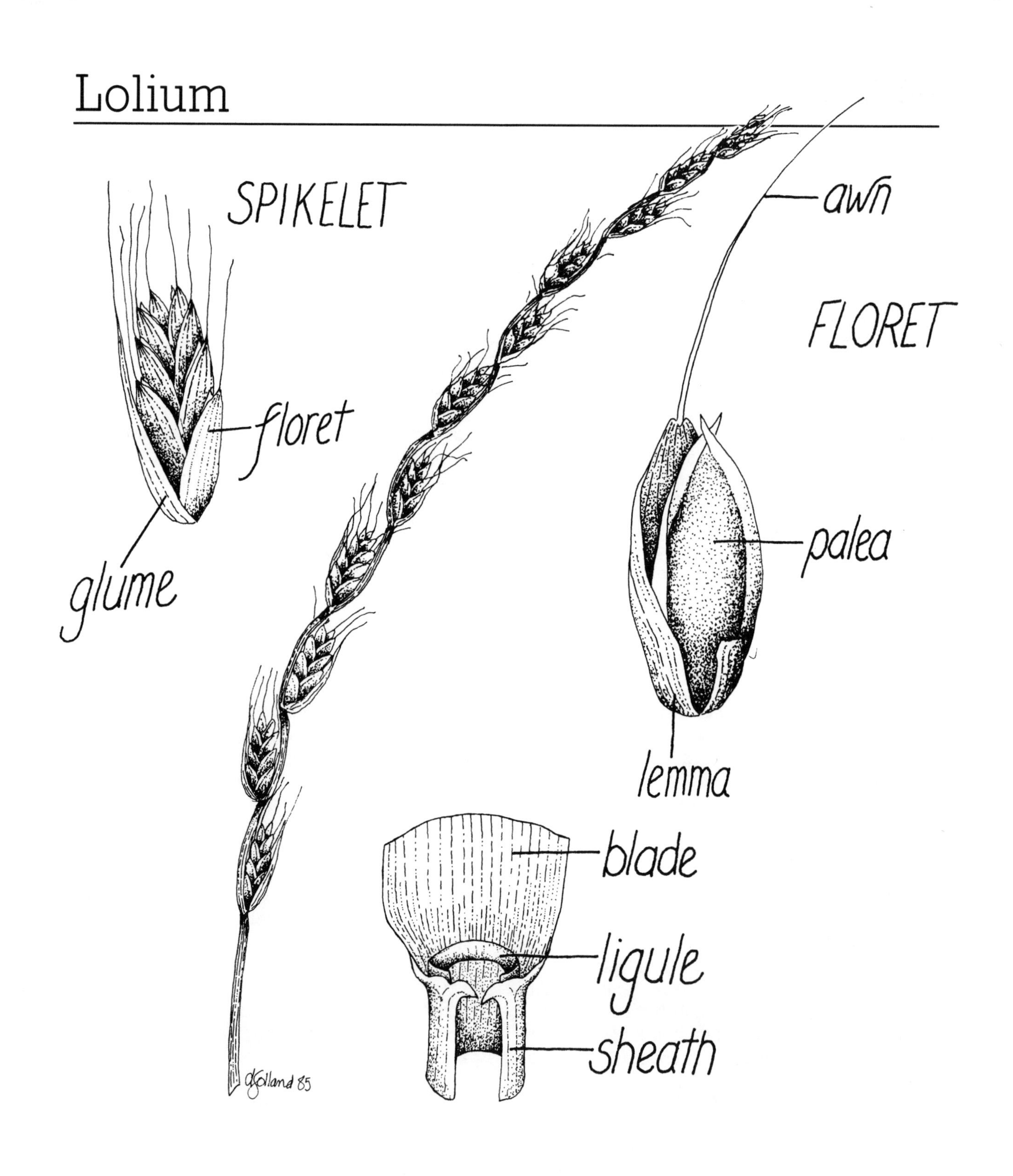


*L. temulentum*

# GENUS: *Lolium*

*Lolium temulentum* L. - Darnel

**FAMILY:** Gramineae (Poaceae) — the Grass Family

This huge, economically important group contains several hundred genera and many thousand species. All of the world's food grains belong to this family, as do bamboo, lawn and turf species, sugar cane, and many other plants vital to humans. Very few species are poisonous. Foxtail grass (*Setaria* spp.) and squirreltail (*Hordeum jubatum* L.) have hard, sharp, floral parts that produce mechanical damage to the eyes, mouth, and digestive system of livestock; others host parasitic fungi that produce toxins (see *Claviceps*).

Flowers in the grass family are highly modified and require special terminology. The basic reproductive unit is a **floret,** consisting of a flower having either male or female (or commonly both) parts, the stamens and ovary, and scales, the **lodicules.** In addition to the flower, the floret contains **lemmas** (bracts) and **palea** (a bract). The collection of florets constitutes a **spikelet** that can easily be recognized by one, or more often two, bracts called the **glumes.** The grass leaf is modified into leaf sheath, ligule, and blade. The **sheath** is the base of a grass leaf originating from a stem node. The **blade** is the flat, foliar, free portion of the leaf. The **ligule** is an outgrowth between the leaf sheath and blade. The vast diversity in size, shape, and arrangement of the reproductive and vegetative organs is used to distinguish grass species. Grass characteristics include, the **flower:** 3-merous; **leaf veins:** parallel; **stamens:** (1-) 3 (-6), separate, filaments slender; **ovary:** superior, 1-celled, 1-seeded; **styles:** (1) 2 (3), usually featherlike; **fruit:** achene, an indehiscent seedlike structure sometimes permanently enclosed between the lemma and palea.

**PHENOLOGY:** *Lolium temulentum* flowers June through August.

**DISTRIBUTION:** This weed of grain fields and waste places is uncommon in Pennsylvania. Distribution records indicate sporadic occurrence in the southeastern corner of the state.

**PLANT CHARACTERISTICS:** Darnel is an annual grass with **stems:** solitary or a few clumped together, 4-8 dm tall; **blades:** glabrous beneath, scabrous above, 3-8 mm wide; **spike:** 1-2 dm; **spikelets:** placed edgewise to the rachis, 5-8 flowered; **glume:** firm, straight, 5-7 nerved, equalling or surpassing the uppermost lemma, 12-22 mm; **lemmas:** obtuse, awned, or awnless.

**POISONOUS PARTS:** The seeds and seed heads are considered poisonous.

**SYMPTOMS:** In humans, darnel poisoning is characterized by the sensation of intoxication, ataxia, giddiness, apathy, various abnormal sensations, mydriasis, nausea, vomiting, gastroenteritis, and diarrhea. It is rarely fatal.

**POISONOUS PRINCIPLES:** The alkaloids temuline and loliine possibly are responsible for toxicity of darnel. It also has been suggested that toxicity may be due to a parasitic fungus living within the seed head.

**CONFUSED TAXA:** The nontechnical characters in the illustration should readily distinguish darnel from the numerous other grass species found in Pennsylvania. The linear leaves have narrow horns (auricles) at the sheath-blade junction; the ligule is membranous and truncate; the spike is rigid, 5-40 cm long, and composed of 5-15 spikelets each 8-26 mm long.

**SPECIES OF ANIMALS AFFECTED:** Though cases of darnel poisoning are rare, humans and livestock are susceptible.

**TREATMENT:** (11a)(b); (26)

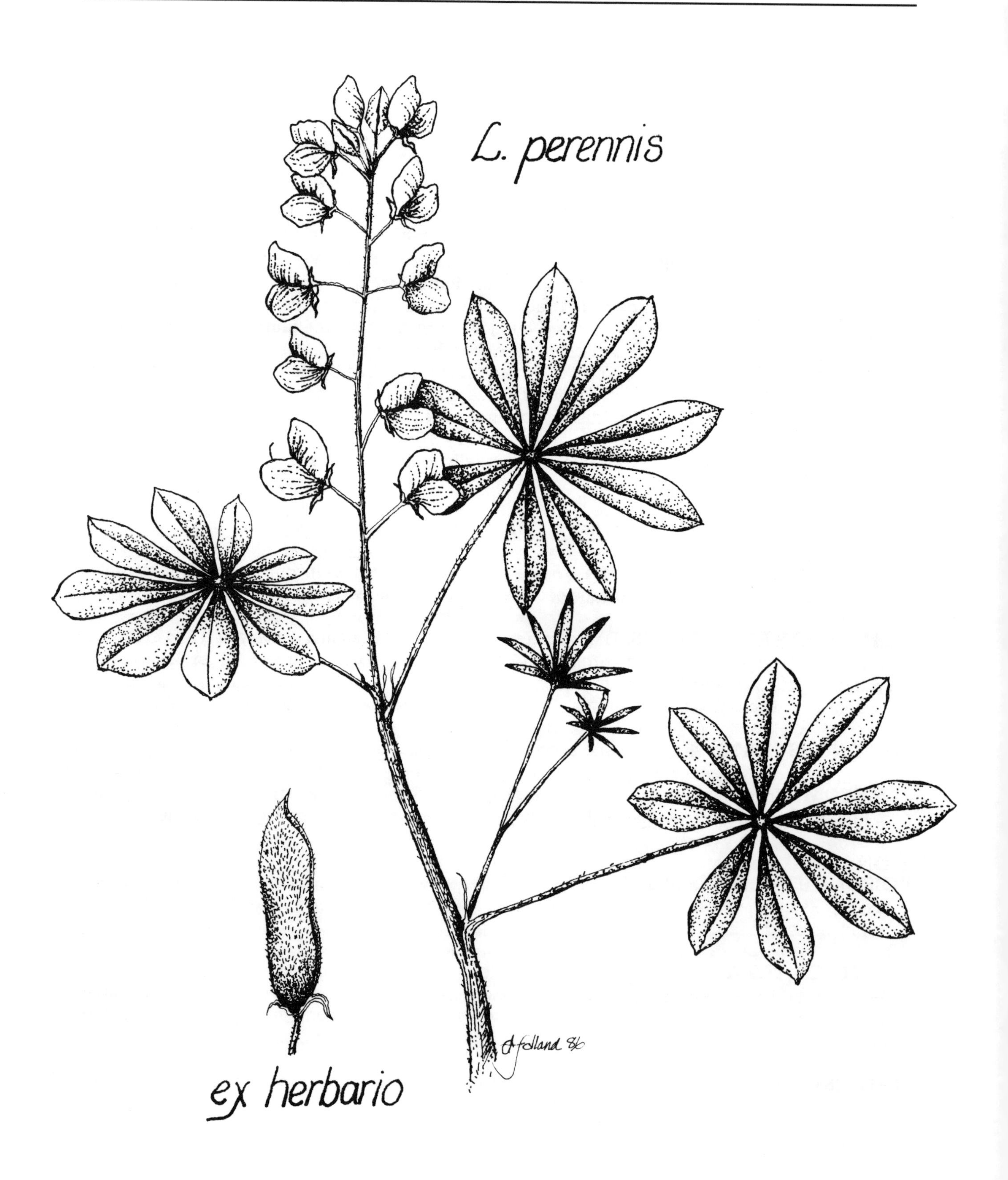
L. perennis
ex herbario
Holland '86

# GENUS: *Lupinus*

*Lupinus perennis* L. — Wild lupine

**FAMILY:** Fabaceae (Leguminosae) — the Bean Family (see *Crotalaria*)

**PHENOLOGY:** *Lupinus perennis* flowers in late spring from April into July.

**DISTRIBUTION:** Wild lupine is found in Pennsylvania in a diversity of habitats ranging from dry open woods and clearings to moist sandy soil.

**PLANT CHARACTERISTICS:** Wild lupine is an erect, perennial **shrub:** 2-6 dm tall, thinly pubescent; **leaves:** palmately lobed; **lower leaves:** 5 cm long, 7-11 leaflets; **petioles:** 2-6 cm; **racemes:** erect, 1-2 dm, numerous, blue varying to pink or white flowers; **flowers:** 2-lipped; **calyx:** the upper lip, 4 mm, 2-toothed; the lower entire, 8 mm; **corolla:** standard, 12-16 mm, half as wide; wings united toward the summit; **stamens:** 10, monadelphous; filaments forming a closed tube for half their length; **pod:** pubescent, 3-5 cm long, oblong, flattened.

**POISONOUS PARTS:** The foliage and seeds are considered poisonous. The vast literature on toxicity of *Lupinus* spp. mainly involves western taxa, e.g. rangeland species. Toxicity may vary among species, produce different symptoms in various classes of livestock, and fluctuate according to season and habitat.

**SYMPTOMS:** The reactions to ingestion are paradoxical. Some animals show depression, others excitation. Respiratory problems generally develop with labored breathing, coma or convulsions, and death.

**Postmortem:** No distinctive lesions are seen in American cases of lupine poisoning. Pregnant cows, pastured in areas of lupine growth, can bear calves afflicted with arthrogryposis, scoliosis, torticollis, and cleft palate. **Postmortem** in mycotoxin-induced European lupinosis shows **gross lesions:** signs of cirrhosis, liver and kidney degeneration, pulmonary edema, congestion of internal organs and gastrointestinal irritation; **histological lesions:** focal centrilobular and midzonal hepatocellular necrosis; hyperplasia of Kupffer cells and bile ductules; and hepatic necrosis and cirrhosis.

**POISONOUS PRINCIPLES:** The majority of the more than 20 alkaloids isolated from *Lupinus* are quinolizidine alkaloids with some piperidine and other components known. Lupanine and lupinine are well-studied compounds from *Lupinus;* spateine appears less well characterized. In Europe, the disease called lupinosis is attributable to mycotoxins produced by the fungus *Phomopsis leptostomiformis,* which grows on *Lupinus* species.

**CONFUSED TAXA:** Several species are cultivated for garden purposes, and many highly ornamental lupines have been developed through hybridization and selection. The "Russell" group, of uncertain parentage, is potentially toxic. The uninitiated gardener might confuse larkspur or delphinium (see *Delphinium*) with lupines. The three are distinct, however, as both larkspur and delphinium have spurred floral parts not present in lupines.

**SPECIES OF ANIMALS AFFECTED:** In western rangeland sheep are more commonly poisoned by lupines than horses or cattle. No uncontested cases of lupine poisoning are known from Pennsylvania.

**TREATMENT:** (11a)(b); (26)

**OF INTEREST:** Because alkaloids remain toxic in dried plants, contaminated hay also is poisonous. Alkaloids are generally more concentrated in plants after flowering, perhaps due to higher concentrations in the seeds.

# Menispermum

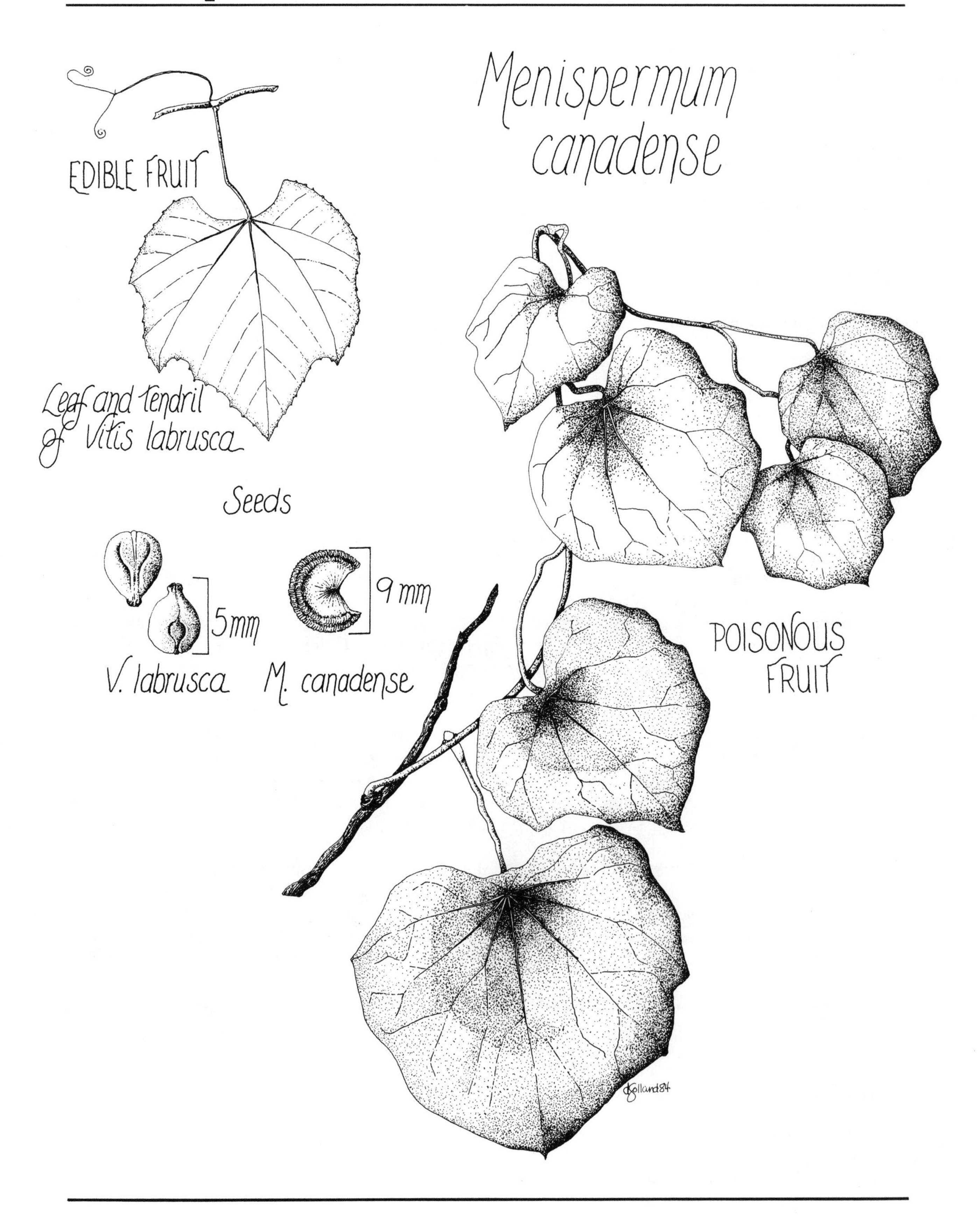

# GENUS: *Menispermum*

*Menispermum canadense* L. — Moonseed

---

**FAMILY:** Menispermaceae — the Moonseed Family

This mostly tropical group is not well represented in Pennsylvania. It includes twining, dioecious **vines** with **leaves:** simple, alternate, lacking stipules, palmately veined; **flowers:** small, inconspicuous, white or greenish, unisexual; **sepals:** not much differentiated from the petals, the outer series **(calyx)** longer than the inner series **(petals); stamens:** 6 (or more); **ovaries:** 3, separate; **fruit:** a drupe.

**PHENOLOGY:** Moonseed flowers June through July.

**DISTRIBUTION:** Found twining on other vegetation in moist woods, thickets, and fencerows.

**PLANT CHARACTERISTICS:** Moonseed is a perennial woody climber, **sepals:** 4 to 8, longer than the 4 to 8 petals; **male flowers:** stamens 12 to 24; **female flowers:** with 2 to 4 pistils; **fruit:** a black drupe, with a whitish wax film at maturity; appearing in grapelike clusters; **leaves:** palmately veined with low, rounded teeth; **leaf petioles:** may be twisted at the point where they attach to the stem.

**POISONOUS PARTS:** Perhaps all parts of this plant are toxic, but the fruits, which hang in pendulous grapelike clusters, are especially poisonous. The rootstocks also contain bitter alkaloids.

**SYMPTOMS:** The symptoms of poisoning could not be found in the literature; death, however, has been reported from ingestion of seeds.

**POISONOUS PRINCIPLES:** Plants contain isoquinoline alkaloids, including dauricine that has curarelike activity. In fact, Amerindians use a South American member of this family, *Chondodendron tomentosum* Ruiz & Pav., as an ingredient in arrow poison. It is also a constituent in a muscle relaxer used by anesthetists prior to operations.

**CONFUSED TAXA:** The woody vine, dark clusters of fruit, and similar general leaf shape allow this plant to be confused with wild grape (*Vitis* L.), the fruit of which is edible. Moonseed can be distinguished from wild grape by closely comparing leaves and seeds contained in the fruit. Wild grape leaves have twenty or more blunt teeth, but moonseed leaves have fewer than ten broad, shallow, rounded lobes. Grape has several slightly pear-shaped seeds within each fruit; moonseed has a single, flat, crescent-shaped (almost circular), grooved seed.

**SPECIES OF ANIMALS AFFECTED:** Birds appear to safely consume the fruits of moonseed plants. Because its distribution is limited to woodlands, *Menispermum canadense* is unlikely to be a problem for livestock. Records indicate that some time before 1935 children in Cambria County, PA, mistakingly ate moonseed fruits and died.

**TREATMENT:** (11a) (b); (26)

# Mirabilis

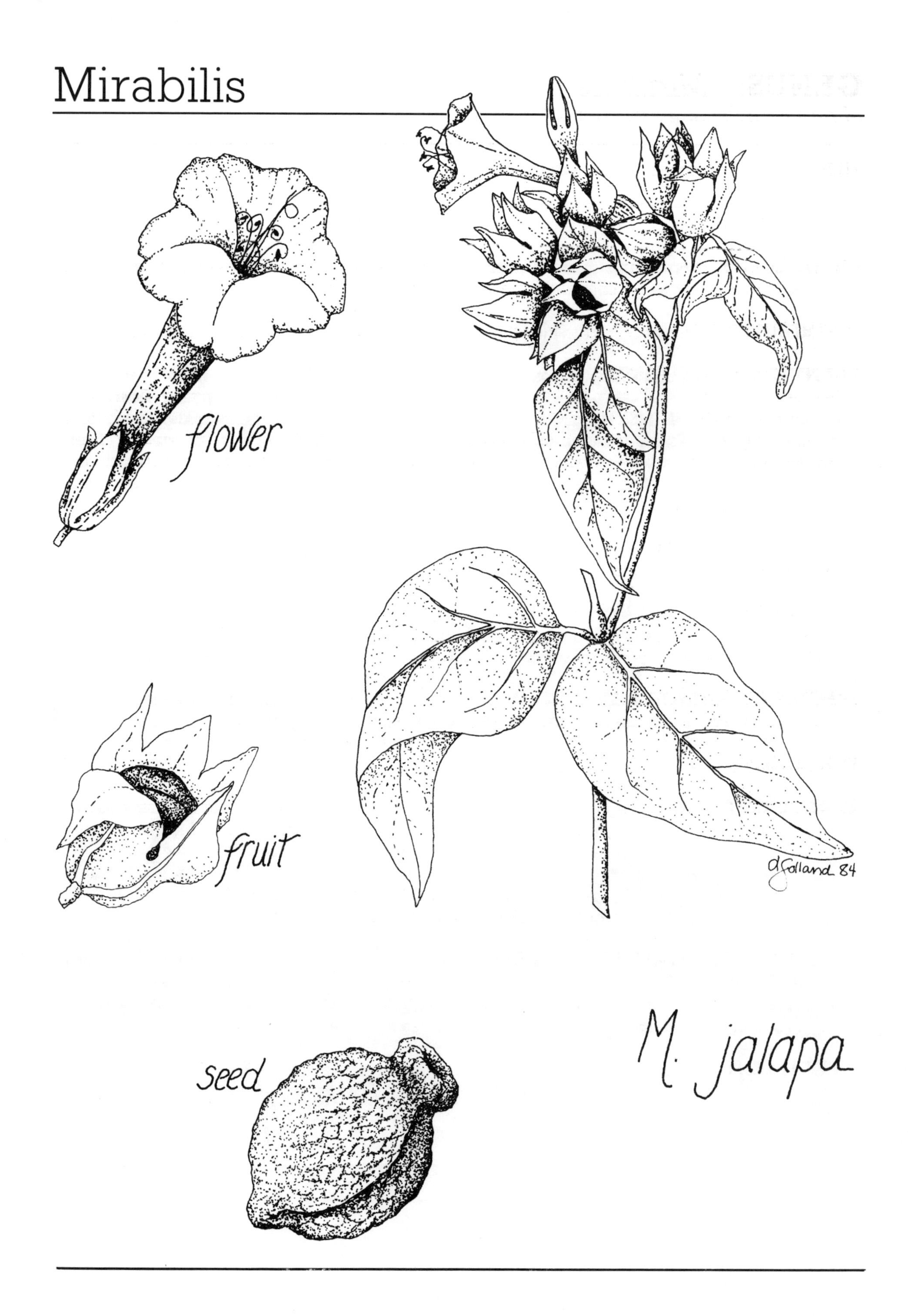

# GENUS: *Mirabilis*

*Mirabilis jalapa* L. — Four o'clock

---

**FAMILY:** Nyctaginaceae — The Four o'clock Family

Relatively few members of this family occur in Pennsylvania. A full description of the poisonous plant is given below.

**PHENOLOGY:** This perennial plant of warmer regions is cultivated in Pennsylvania as a garden annual. It flowers in the summer and is fragrant in the evening hours.

**DISTRIBUTION:** This widely cultivated plant occasionally escapes to roadsides and wasteplaces.

**PLANT CHARACTERISTICS:** Four-o'clock is a much-branched, erect plant growing to 1 m tall; **leaves:** ovate, opposite, deep green, the lower, primary ones petioled, acuminate; **flowers:** 3.5 cm across, red, pink, yellow, or white, often striped and mottled, opening in late afternoon or during daytime in cloudy weather; **calyx tube:** corollalike, 2-3 cm, 5-lobed at top; **corolla:** absent; **involucre:** 5-lobed, calyxlike, 6-8 mm at flowering; **fruit:** rounded at the summit, tapered at base.

**POISONOUS PARTS:** The roots and seeds are reported to be poisonous; herbage should also be considered suspect.

**SYMPTOMS:** Gastroenteritis, including vomiting, diarrhea, and abdominal pain, are symptomatic.

**POISONOUS PRINCIPLES:** The toxic principles are unknown.

**CONFUSED TAXA:** No plant, native or introduced, is readily confused with four o'clock.

**SPECIES OF ANIMALS AFFECTED:** Children have been poisoned from ingestion of four-o'clock roots and seeds. Livestock might also be susceptible.

**TREATMENT:** (11a)(b); (26)

**OF INTEREST:** *Mirabilis jalapa* has shown some potential as an anticancer chemotherapeutic.

# Narcissus

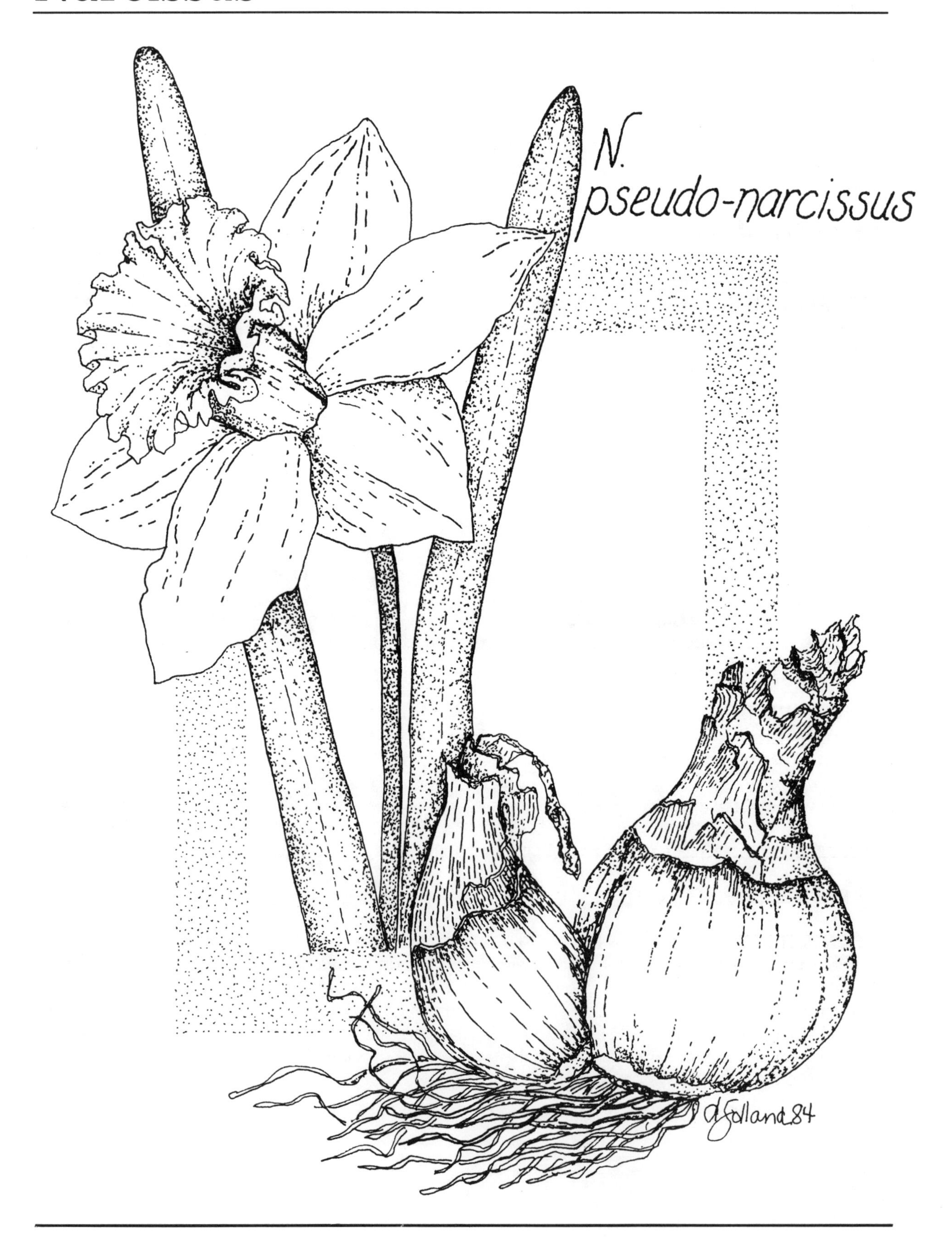

# GENUS: *Narcissus*

*Narcissus Pseudo — Narcissus* L. — Daffodil

---

**FAMILY:** Amaryllidaceae — the Amaryllis Family

Only one genus of this worldwide family, *Hypoxis,* is native to the Commonwealth. Other amaryllis encountered in Pennsylvania, such as narcissus, daffodil, snowdrops, and snowflakes, have escaped from European introductions. **Flowers** are generally bisexual and regular, bearing 6 perianth parts in 2 series; in some *Narcissus* a crown or **corona** is present; **ovary:** single, with 1 pistil; **stamens:** 6; **fruit:** a trilocular capsule.

**PHENOLOGY:** Daffodils are one of the earliest spring-flowering plants. In protected spots the bright yellow flowers appear in March and April.

**DISTRIBUTION:** *Narcissus* often escape from cultivation and may be encountered in dense colonies along roadsides, moist meadows, and clearings in the woods. Occasionally old homesteads, the houses no longer standing, can be identified by a row of daffodils in what is now woods.

**PLANT CHARACTERISTICS: Floral scapes:** 2-4 dm, nearly equalling the linear, parallel-veined leaves; **flowers:** yellow, solitary, 4-6 cm wide; crown as long as the tepals, often frilled.

**POISONOUS PARTS:** All parts are poisonous, especially the bulbs.

**SYMPTOMS:** Severe gastroenteritis, vomition, purging, nervous symptoms such as trembling and convulsions, diarrhea, nausea, and death can result from bulb consumption. Irritant dermatitis also can occur when the needle-sharp calcium oxalate crystals, distributed in the outer layers of many Narcissus bulbs, pierce the hands of those working with them. The "wheals" are characteristic of the disease "bulb fingers," a symptom suggestive of histamine release.

**POISONOUS PRINCIPLES:** Active principles that cause poisonings are unknown.

**CONFUSED TAXA:** Several species of *Narcissus* are cultivated; the more common taxon is the daffodil listed above. However, the poet's narcissus, *N. poeticus* with a smaller white perianth and short (a fourth as long as the tepals) yellow/red-margined corona, also escapes. *Narcissus incomparabilis* Mill. (corona half as long as the tepals) and *N. Jonquilla* L. with 2-6 yellow flowers (2-4 cm wide) per scape also are encountered here.

**SPECIES OF ANIMALS AFFECTED:** Small amounts of the bulb have caused human poisoning.

**TREATMENT:** (11a)(b); (26)

**OF INTEREST:** Some commonly cultivated members of the amaryllis family that contain alkaloids known to poison livestock include *Amaryllis* spp. and *Galanthus nivalis* L. (snowdrops). In the Netherlands cases of poisoning occurred when the bulbs were fed to livestock as emergency feeds during World War II. Snowflakes (*Leucojum aestivum* L.), also cultivated and escaped in our range, should be treated with suspicion. Snowflakes produce galanthamine (as do some members of the genus *Galanthus*), which is used medicinally in Europe to treat myasthenia gravis, a muscle and somatic nervous system disorder.

# Ornithogallum

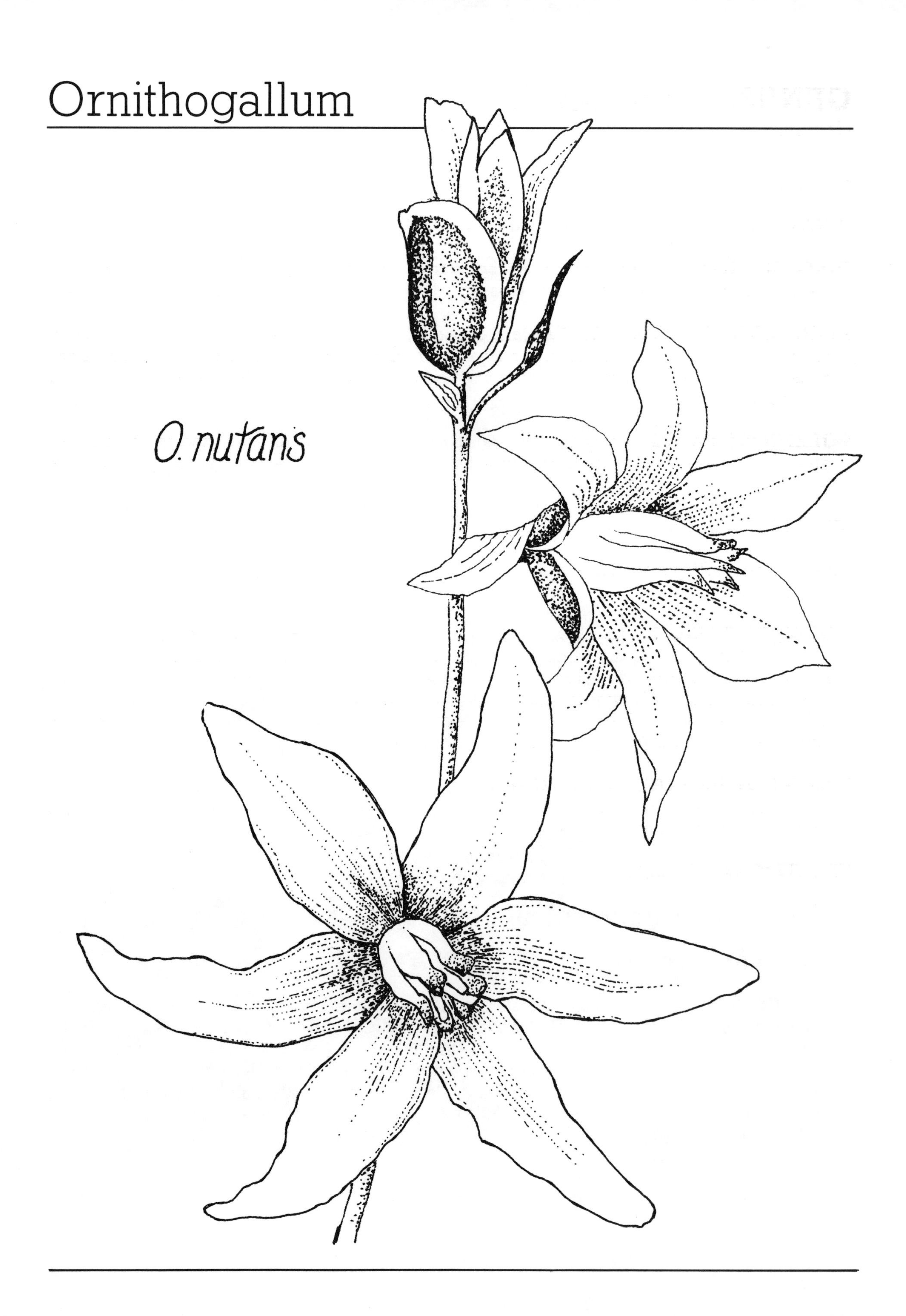

# GENUS: *Ornithogallum*

*Ornithogallum umbellatum* L. — Star-of-Bethlehem

---

**FAMILY:** Liliaceae — the Lily Family (see *Amianthium*)

**PHENOLOGY:** Star-of-Bethlehem flowers in May and June.

**DISTRIBUTION:** This species has escaped from cultivation into roadsides, meadows, and clearings in woods. It has been planted as a garden ornamental but is used less frequently now.

**PLANT CHARACTERISTICS: Tepals:** 6, separate, white with a broad green midstripe beneath; **stamens:** 6; perennial **herbs** from a truncated bulb; **leaves:** linear, basal, 2-4 mm wide; **flowering stems:** leafless, star-shaped flowers are subtended by a small bract; **fruit:** a trilobed, several-seeded capsule.

**POISONOUS PARTS:** Although the aerial portions of the plant are reported to contain toxic alkaloids, animals seem to graze on them without adverse effect. The small, white, onionlike bulb is toxic.

**SYMPTOMS:** Ingestion causes nausea, gastrointestinal upset, bloating, depression, and salivation.

**POISONOUS PRINCIPLES:** An unidentified alkaloid, perhaps closely related to colchicine, is responsible for toxicosis.

**CONFUSED TAXA:** Two species of *Ornithogallum* occur in the Commonwealth: *O. umbellatum* and *O. nutans* L. The literature reports *O. umbellatum* as toxic; in the absence of information, *O. nutans* also should be considered poisonous. The following characteristics differentiate the two: *Ornithogallum umbellatum* flowers in May-June, has narrow leaves 2-4 mm wide, and an inflorescence in which the lower flowers are on long ascending pedicels; *O. nutans* flowers in April-May, has leaves 4-8 mm wide, and equal pedicels.

**SPECIES OF ANIMALS AFFECTED:** Children have been poisoned from consuming flowers and bulbs. Sheep and cattle have died from eating bulbs. Frost heaving of soil and plowing will bring the bulbs to the soil surface.

**TREATMENT:** (11a)(b); (26)

# Parthenocissus

*P. quinquefolia*

# GENUS: *Parthenocissus*

*Parthenocissus quinquefolia* (L.) Planch. — Virginia creeper
*Parthenocissus vitacea* (Knerr) Hitchc. — Virginia creeper

---

**FAMILY:** Vitaceae - the Grape Family

This family consists of **woody vines** climbing by tendrils; **flowers:** 4- or 5-merous; **ovary:** superior, 2-6 celled, surrounded by a glandular disk; **fruit:** a berry.

**PHENOLOGY:** Both species of Virginia creeper flower in June.

**DISTRIBUTION:** Virginia creeper is distributed on moist soil throughout northeastern United States.

**PLANT CHARACTERISTICS:** The vines can be recognized by the alternate leaf arrangement; **leaves:** palmately compound with 5 leaflets; **fruit:** a black berry.

**POISONOUS PARTS:** The berries are suspected to be toxic.

**SYMPTOMS:** The toxicology of this genus is not well studied although poisoning and death in humans has been suggested in the literature. Possible symptoms include vomiting, diarrhea, and a feeling that bladder or intestinal discharge should occur.

**POISONOUS PRINCIPLES:** The toxic principles are unknown.

**CONFUSED TAXA:** The two species may be distinguished by the numerous adhesive disks on the many-branched tendrils of *P. quinquefolia*. The occurrence of adhesive disks on the few-branched tendrils of *P. vitacea* is rare. Boston or Japanese ivy, *P. tricuspidata* (Siebold & Zucc.) Planch., a cultivated vine grown around dwellings, has simple, glossy, 3-lobed or 3-parted leaves. Poison ivy (see *Rhus radicans*) has compound leaves with 3 leaflets and is often confused with Virginia creeper.

**SPECIES OF ANIMALS AFFECTED:** The berries have been implicated in the death of children. In one feeding study, twelve berries were deadly to a guinea pig.

**TREATMENT:** (11a)(b); (26)

# Philodendron

# GENUS: *Philodendron*

*Philodendron* — Philodendron

**FAMILY:** Araceae — the Arum Family (see *Arisaema*)

**PHENOLOGY:** Like *Dieffenbachia* and other aroids, philodendrons are cultivated primarily for their lush green growth; they seldom flower in cultivation.

**DISTRIBUTION:** Philodendron, a very common house plant grown for its foliage, is native to the warm regions of the Americas, including the West Indies. The genus has 200 species.

**PLANT CHARACTERISTICS:** Philodendrons vary considerably in appearance. Identification of the many species is complicated by differences in form and size of leaves between juvenile and adult growth stages, and by the vast number of hybrids that have been produced commercially. Young plants of many species have similar leaves and often are impossible to identify. Most philodendrons have climbing **stems** with aerial roots, although some may be erect and free-standing; **plants:** evergreen, perennial; **leaves:** alternate, large, thick and shining, entire to variously lobed or pinnatifid. The philodendron variations are too numerous to list in this brief treatment.

**POISONOUS PARTS:** All parts of the philodendron plant are toxic. Leaves and stems are dangerous when eaten in quantity.

**SYMPTOMS:** In addition to those symptoms described for *Arisaema* and *Dieffenbachia,* philodendron can cause mouth, tongue, and lip irritation. One researcher has reported 72 cases of "philodendrum" (sic) poisoning in cats, with more than half resulting in deaths. Symptoms included debilitation, listlessness, and kidney malfunction, although these were not associated with apparent pain.

**POISONOUS PRINCIPLES:** Aroid toxins, including calcium oxalate needles and perhaps proteins or amino acids, are responsible for toxicosis.

**CONFUSED TAXA:** The philodendrons can be informally grouped into four categories: trailing with slender, weak stems; stouter-stemmed vines with entire leaves; stouter-stemmed vines with lobed ("cut-leaf") leaves; and shrubby, nonvining ("giant"). Of the trailing type, *P. scandens* C. Koch & H. Sello and *P. cordatum* (Vello) Kunth are commonly grown. Within the stouter-stemmed vine (entire leaves) category, *P. domesticum* Bunt is popular. Those stout vines with divided leaves include *P. radiatum* Schott and hybrids like *P.* x *'Florida.'* Shrubby philodendrons generally are not encountered in homes in Pennsylvania; however, they are planted in tropical gardens, grown under glass in conservatories in the temperate region, and used in interior displays at shopping malls.

Other genera sometimes confused with vining philodendron are *Pothos* and *Scindapsus;* some "split-leaf" philodendron are actually *Monstera* and *Epipremnum.* Because the genera listed above also are in the Araceae, they should be considered potentially dangerous.

**SPECIES OF ANIMALS AFFECTED:** Humans and house pets, especially cats, are susceptible to arum toxins.

**TREATMENTS:** Aroids, including *Arisaema, Dieffenbachia,* and *Philodendron,* can be treated similarly. General treatment includes: (6); (2 - diazepan i.v.); (11a) (b) except in severe swelling; milk, water, or antacids to dilute the calcium oxalate and to flush out and soothe the oral pharynx; analgesics (e.g. meperidine); (4 - effectiveness is equivocal); and maintenance of hydration (intravenous fluids).

# Phoradendron

# GENUS: *Phoradendron*

*Phoradendron serotinum* (Raf.) M. C. Johnst. — Mistletoe

**FAMILY:** Loranthaceae. — the Mistletoe Family

This family is characterized by semiparasitic plants, attached to trees or shrubs by **haustoria,** lacking ordinary roots, but having green (chlorophyllous) leaves and stems; **leaves:** opposite; **flowers:** inconspicuous; **ovary:** inferior; **stamens:** as many as and opposite the perianth-lobes.

**PHENOLOGY:** *Phoradendron serotinum* flowers from May through July.

**DISTRIBUTION:** An uncommon semiparasite, mistletoe can be found parasitizing several different species of deciduous trees in Pennsylvania woods. It can be purchased from novelty shops and grocery stores at Christmas time.

**PLANT CHARACTERISTICS:** This species of mistletoe appears as small **shrubs,** parasitic on trees; **leaves:** coriaceous, opposite, entire, oblong to obovate, 2-6 cm, blunt or rounded; **stems:** freely branching, thick, brittle; **calyx:** deeply 3-lobed; **anthers:** 3, sessile on the base of the calyx-lobes; **ovary:** ovoid, with 1 subsessile stigma; **flowers:** small, in short axillary spikes; **fruit:** a white, mucilaginous berry.

**POISONOUS PARTS:** The berries are especially poisonous, but leaves and stems also are toxic.

**SYMPTOMS:** Gastrointestinal irritation, diarrhea, weakened pulse, and cardiovascular collapse have been reported in humans. Cattle have died after consuming mistletoe but have shown no symptoms or significant lesions.

**POISONOUS PRINCIPLES:** The toxins are pressor amines, beta-phenylethylamine and tyramine.

**CONFUSED TAXA:** The familiar Christmas mistletoe has no readily confused counterparts. The nomenclature of this species has been in question; some authors cite the plant as *P. flavescens* (Pursh) Nutt.

**SPECIES OF ANIMALS AFFECTED:** Humans and cattle are susceptible to the amines in mistletoe. Other animals, including livestock and house pets, should not be allowed to eat the leaves or berries.

**TREATMENT:** (11a)(b); (26); (19)

**OF INTEREST:** For some individuals the plant can be an irritant or cause dermatitis upon contact. The alkaloid tyramine, found in mistletoe, is a vasopressor, elevating blood pressure; it is classed as an adrenergic chemical. The American Indians used *P. serotinum* extracts to stop postpartum hemorrhage.

# Phytolacca

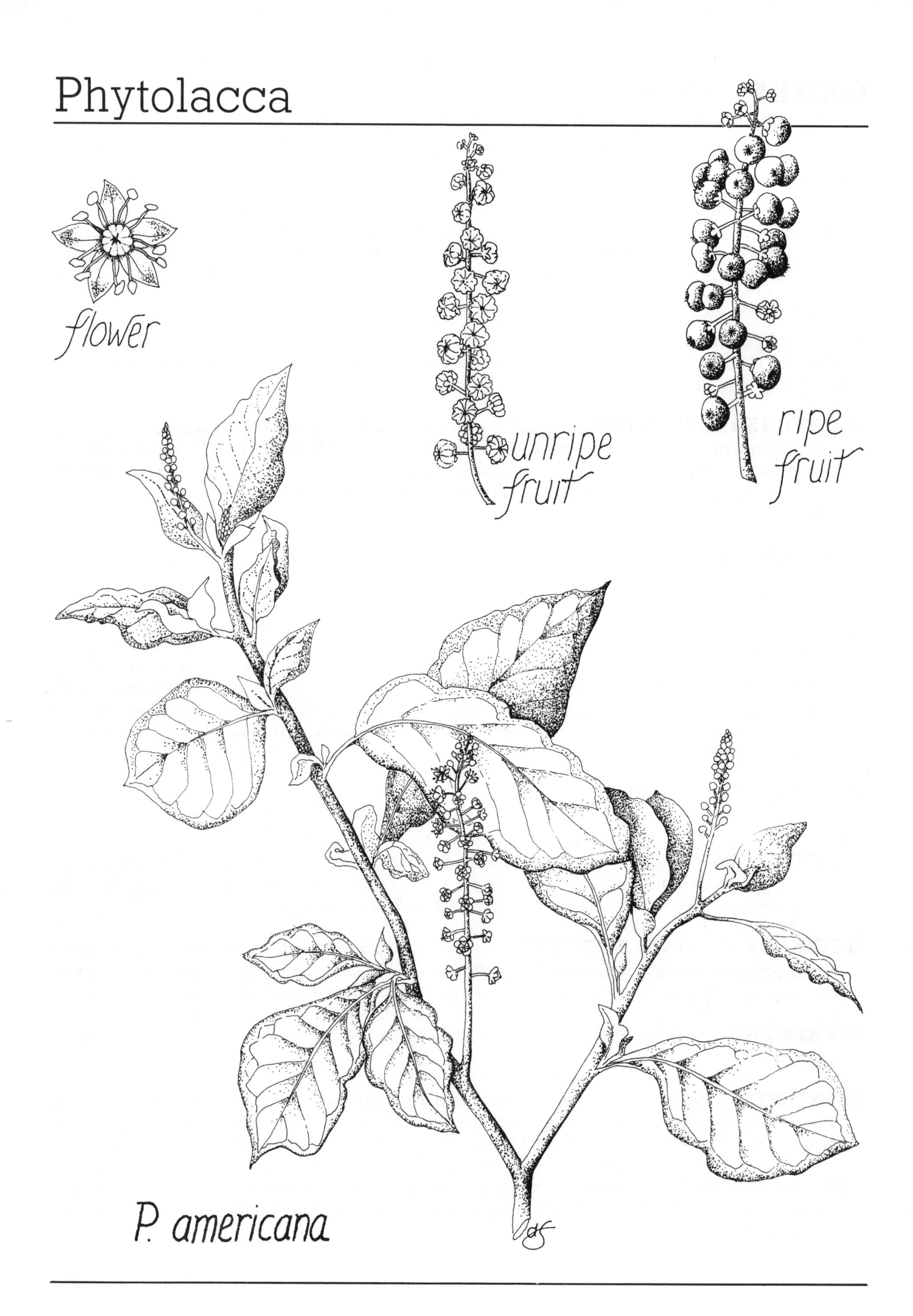

# GENUS: *Phytolacca*

*Phytolacca americana* L. — Pokeweed; pokeberry; inkberry

**FAMILY:** Phytolaccaceae — the Pokeweed Family

Native to both America and Africa, this family is composed of plants with **leaves:** alternate, entire; **flowers:** in racemes, bisexual (or unisexual); **calyx:** 4- to 5- parted; **petals:** absent; **stamens:** 3 to many; **ovary:** superior (or partly inferior); **fruit:** drupelike berries. *Phytolacca americana* is the only plant of the Phytolaccaceae found in Pennsylvania.

**PHENOLOGY:** Pokeweed flowers July through September.

**DISTRIBUTION:** Found in rich, disturbed soils such as barnyards, lowlands, fields, fencerows, and moist woodland.

**PLANT CHARACTERISTICS:** *Phytolacca americana* can be identified by **sepals:** greenish white to pink; **flowers:** 6 mm wide; **racemes:** 1-2 dm, pedunculate; **infructescence:** nodding; **stamens:** 10; **pistils:** 10; **fruit:** 5-15 cells, a 1 cm thick, juicy (inky), shiny, dark-purple berry; **plants:** glabrous, perennial herbs, to 3 m tall, branched above; **leaves:** lance-oblong to ovate, 1-3 dm; **petioles:** 1-5 cm.

**POISONOUS PARTS:** All parts, but primarily the roots, are considered poisonous. Small quantities (more than 10) of raw berries can result in serious poisoning of adults. Fatalities in young children can result from the consumption of a few raw berries.

**SYMPTOMS:** The more common symptoms are gastrointestinal cramps, vomiting, diarrhea, and convulsions in severe cases. Perspiration, prostration, weakened respiration and pulse, salivation, and visual disturbance are possible symptoms. Death may result. Humans experience an immediate burning sensation in the mouth upon consumption. **Postmortem: gross lesions:** mild to severe gastroenteritis; congestion of internal organs; **histological lesions:** stomach ulcerations with hemorrhage.

**POISONOUS PRINCIPLES:** The physiologically active principles have been identified. Suspected compounds include saponin, together with lesser amounts of the alkaloid phytolaccin.

**CONFUSED TAXA:** Few plants are confused with pokeweed. The infructescence may superficially resemble that of chokecherry or wild cherry (see *Prunus*), but *Prunus* is an arborescent plant with woody bark, whereas *Phytolacca* is herbaceous.

**SPECIES OF ANIMALS AFFECTED:** Any class could be affected; however, the plant stem, leaves, and berries are unpalatable and therefore are not usually ingested. Pigs may become ill from routing and eating the roots. Humans may be affected if they eat the berries, stems, or roots.

**TREATMENT:** (11a)(b); (26); peripheral plasmacytosis with potential immunosuppressive properties.

**OF INTEREST:** Cooked, young, tender leaves and stems are eaten by some people as a pot-herb. These young greens are the "poke salad" of Southern fame. They contain low concentrations of phytolacca toxin which is destroyed by proper cooking. Cooked berries are edible and occasionally used in pies. *Phytolacca americana* contains mitogens, compounds that can be absorbed through skin abrasions, causing blood abnormalities. Sensitive individuals should handle pokeweed with gloves. Root preparations have been used as a folk-medicinal, a practice that can be dangerous.

# Podophyllum

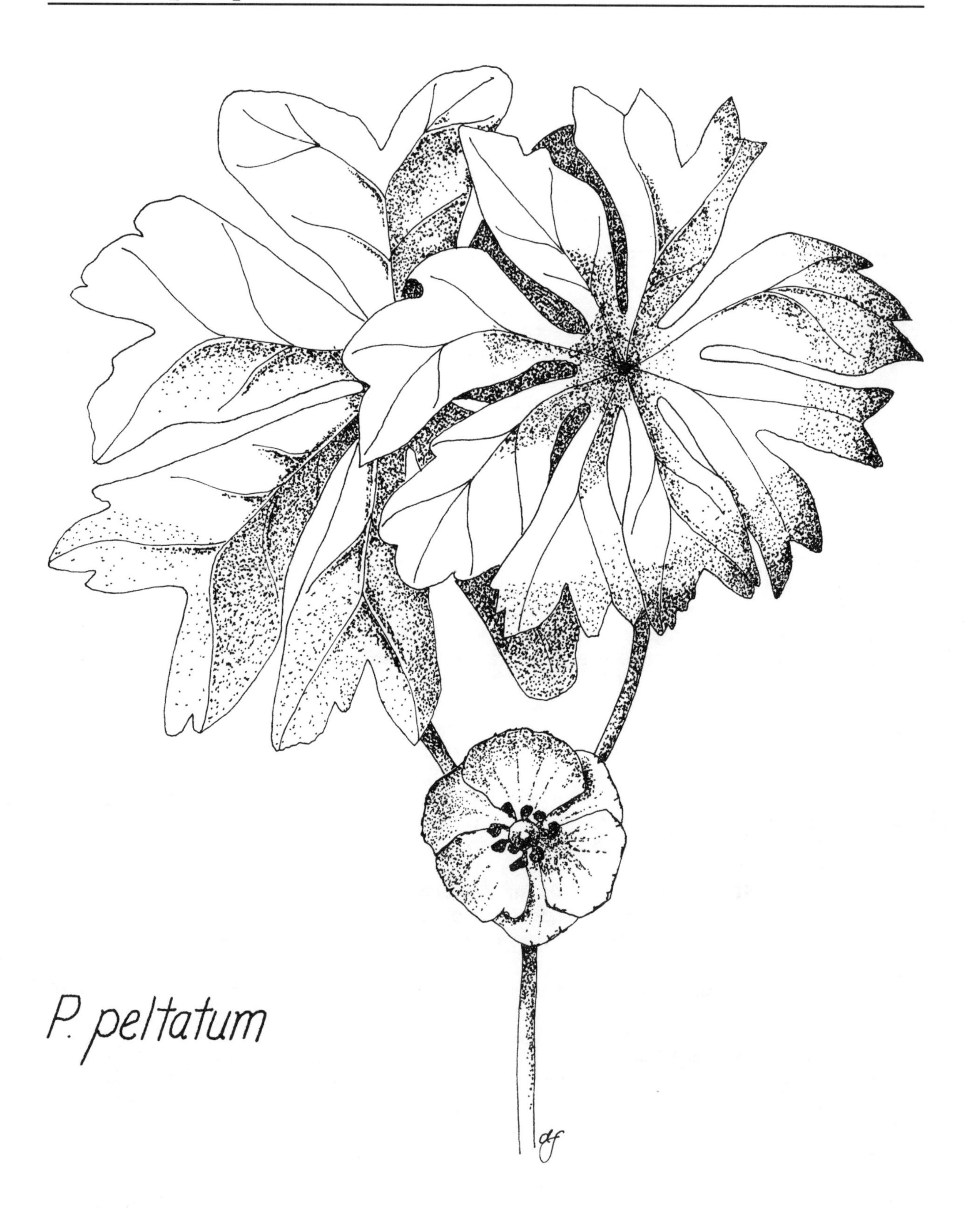

# GENUS: *Podophyllum*

*Podophyllum peltatum* L. — May apple; mandrake

---

**FAMILY:** Berberidaceae — the Barberry Family (see *Caulophyllum*)

**PHENOLOGY:** Mandrake flowers in mid-spring, often during May.

**DISTRIBUTION:** *Podophyllum* is found in open clearings in moist woods and along road banks as a migrant from adjacent wood lots. It is also encountered in wet or damp meadows, open fields, and pastures.

**PLANT CHARACTERISTICS:** *Podophyllum* can be recognized by **sepals:** 6, falling early; **petals:** 6-9, white, 1-2 cm long; **stamens:** twice as many as the petals; **ovary:** oval, with a large sessile stigma; **fruit:** yellow when ripe, 4-5 cm, fleshy pulp edible, many-seeded; **plants:** in colonies; perennial from a rhizome; the flowering stem with two, umbrella-shaped leaves and a short-peduncled, solitary flower in the axil.

**POISONOUS PARTS:** The herbage, rootstock, and seeds are poisonous.

**SYMPTOMS:** In humans and livestock symptoms vary and generally involve severe gastroenteritis, diarrhea, vomiting, and violent catharsis.

**POISONOUS PRINCIPLES:** Podophyllin, a resinoid toxin, is a very complex mixture of lignins (including podophylloxin, alpha- and beta- peltatins) and flavonols. Sixteen physiologically active, well-characterized compounds have been isolated in podophyllin. Chemical analysis reveals 3-6% resin and 0.2 - 1.0% podophyllotoxin, picropodophyllin, quercetin, and peltatins.

**CONFUSED TAXA:** May apples are well known elements of our spring flora. No other plant has umbrellalike leaves and white flowers measuring 5 cm in diameter. It is not readily confused with any other plant.

**SPECIES OF ANIMALS AFFECTED:** Humans, especially adults, have been poisoned from the misuse of medicinal preparations. The fruits, the least toxic part of the plant, have caused poisoning in children. The principal effect is violent diarrhea and vomiting. Where rhizomes are dried and processed at commercial operations, the handlers often show severe conjunctivitis, keratitis, and ulcerative lesions. As little as 5 grains of podophyllotoxin resin can cause death in humans. A cow is known to have been poisoned in Ontario. The animal displayed diarrhea, salivation, anorexia, lacrimation, and excitement; regions of the face and mouth were swollen and the mucosa congested. Other livestock reported poisoned from May apple or mandrake are hogs and sheep.

**TREATMENT:** (11a)(b); (26); (1); (3)

**OF INTEREST:** Preparations from mandake root are commercially available in health food stores. This plant has enjoyed a favored place in homeopathic medicine. Extracts have been used to treat condyloma acuminatum, a type of venereal wart. Prescription preparations still contain podophyllin. Podophyllotoxin is a mitotic poison that kills embryos selectively and is (questionably) teratogenic for surviving fetuses. For this reason pregnant women should avoid the extract of the rootstock. As a mitotic poison, podophyllin and related compounds show tumor-damaging activity and offer some promise in cancer research. *Podophyllum* resin is extremely caustic to the skin and mucous membranes; the resin dust inflames the eyes. It also has been used internally as a purgative in veterinary science.

# Prunus

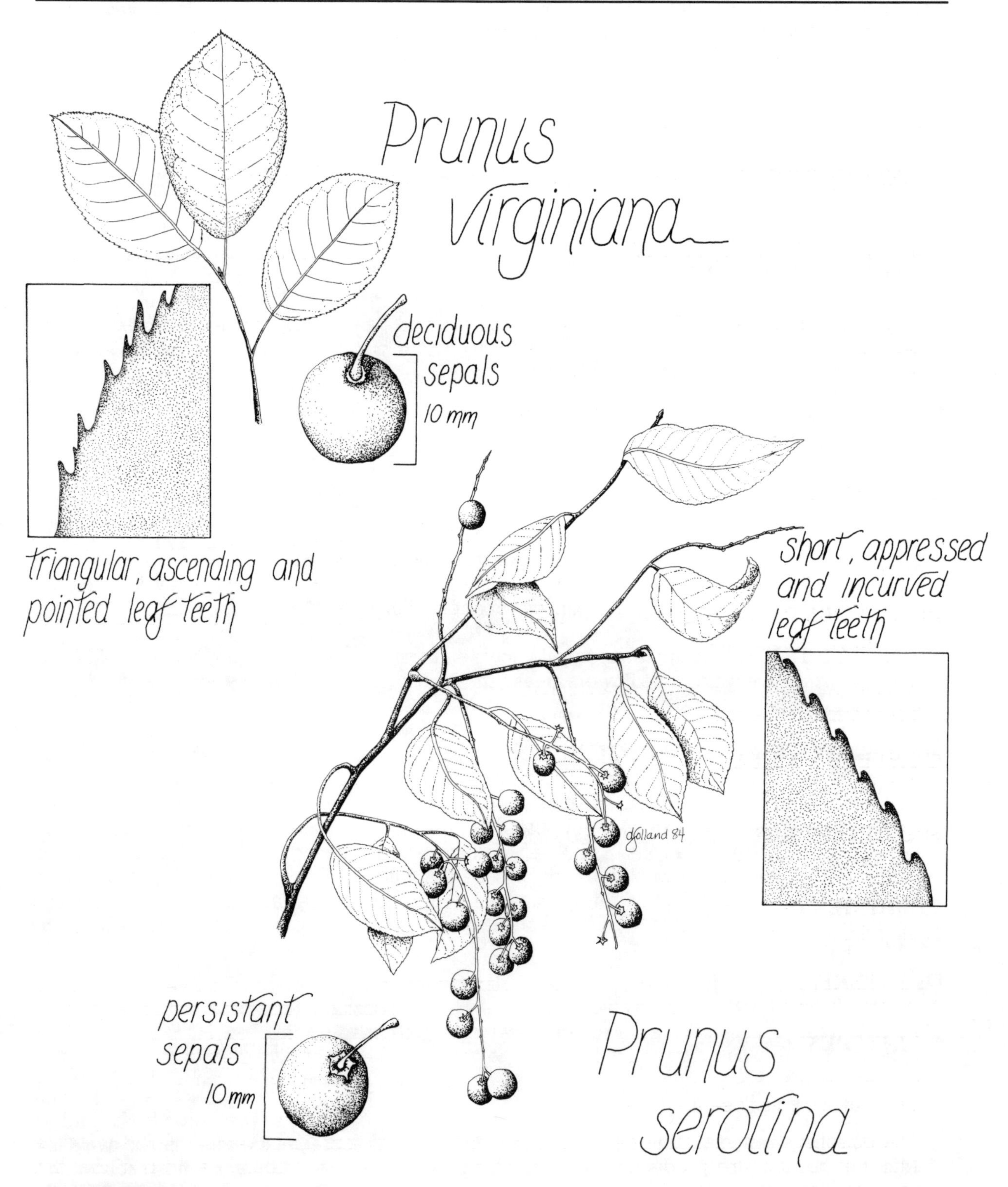


**GENUS:** *Prunus*

*Prunus serotina* Ehrh. — Wild black cherry
*Prunus virginiana* L. — Choke cherry

**FAMILY:** Rosaceae - the Rose Family

This large family of plants contains many genera and numerous species with **leaves:** mostly alternate; **flowers:** bisexual, regular, and 5-merous; **stamens:** 5 to many, borne on the edge of a calyx tube; **pistils:** 1 to many; **ovary:** superior or inferior; **fruits:** achenes, follicles, berries, pomes, or drupes.

**PHENOLOGY:** Both species flower in May.

**DISTRIBUTION:** Both species are found along roadsides, fencerows, waste land, and forest margins.

**PLANT CHARACTERISTICS:** The two species are similar in appearance. *Prunus serotina* is a **tree** to 25 m; **leaves:** lanceolate to oblong or oblanceolate (ovate in *P. virginiana*) and 6-12 cm; acuminate at the tip; serrated with incurving, blunt, callous teeth (sharply serrulate in *P. virginiana*); **racemes:** 8-15 cm long, terminating current-season leafy twigs; **pedicels:** 3-6 mm; **sepals:** 1-1.5 mm long, persistent in *P. serotina,* deciduous in *P. virginiana;* flower **petals:** white and 4 mm long; **fruit:** dark purple or black (red in *P. virginiana*); 1 cm thick. The inner **bark** is aromatic only in *P. serotina.*

**POISONOUS PARTS:** Leaves, twigs, bark, and the stone (pit) produce toxicosis.

**SYMPTOMS:** Poisoning produces anxiety, staggering, falling down, convulsions, dyspnea, rolling of eyes, tongue hanging out of mouth, loss of sensation, dilated pupils; the animal then becomes quiet, bloats, and dies within a few hours of ingestion. **Postmortem: gross and histological lesions:** bright red blood; congestion of internal organs.

**POISONOUS PRINCIPLES:** Cyanogenic glycosides (prunasin, produced in leaves and twigs, and amygdalin, produced in the pit) release hydrocyanic acid (HCN). Less than 1/4 lb of fresh leaves can be toxic to a 100 lb animal. Conflicting reports suggest wilting may increase HCN release. Wilted leaves are more toxic per unit weight due to loss of water by the leaves, which concentrates the cyanide.

**CONFUSED TAXA:** The genus *Prunus* is readily recognized from the description provided. Pokeweed *(Phytolacca americana)* berries may be confused with those of *Prunus* (see *Phytolacca*).

**SPECIES OF ANIMAL AFFECTED:** Humans and all species of livestock are susceptible to HCN poisoning from the cyanogenic glycosides.

**TREATMENT:** (11a)(b); (25); (26); immediate injections of sodium thiosulfate and sodium nitrite may alter the minimum lethal dose.

**OF INTEREST:** *Prunus* contains many useful plants that also may be poisonous. Peach pits (*Prunus persica* Batsch.) are rich in cyanide and have been responsible for animal toxicosis. Apricot kernels (*Prunus armeniaca* L.) have been fatal when consumed by children. Plum pits and bitter almond pits are also cyanogenic. It should be noted that seeds of both the common apple and crabapple (*Malus* spp.) contain HCN. The death of a man, resulting from eating a cup of apple seeds at once, has been reported.

The poisonous ornamental jetbead bush (*Rhodotypos scandens* (Thunb.) Mak. = *R. tetrapetala* Mak.) of horticulture produces a cluster of four, black, shining, berrylike drupes, that are persistent even in winter. The attractive drupes are subtended by 4 spreading, jagged sepals. The shrub grows to 2 m in cultivation. Jetbead has greenish-brown twigs and opposite leaves, glabrous, doubly serrate, to 4 cm long. Symptoms are those for amygdalin poisoning discussed above; treatment; (11a)(b);(25).

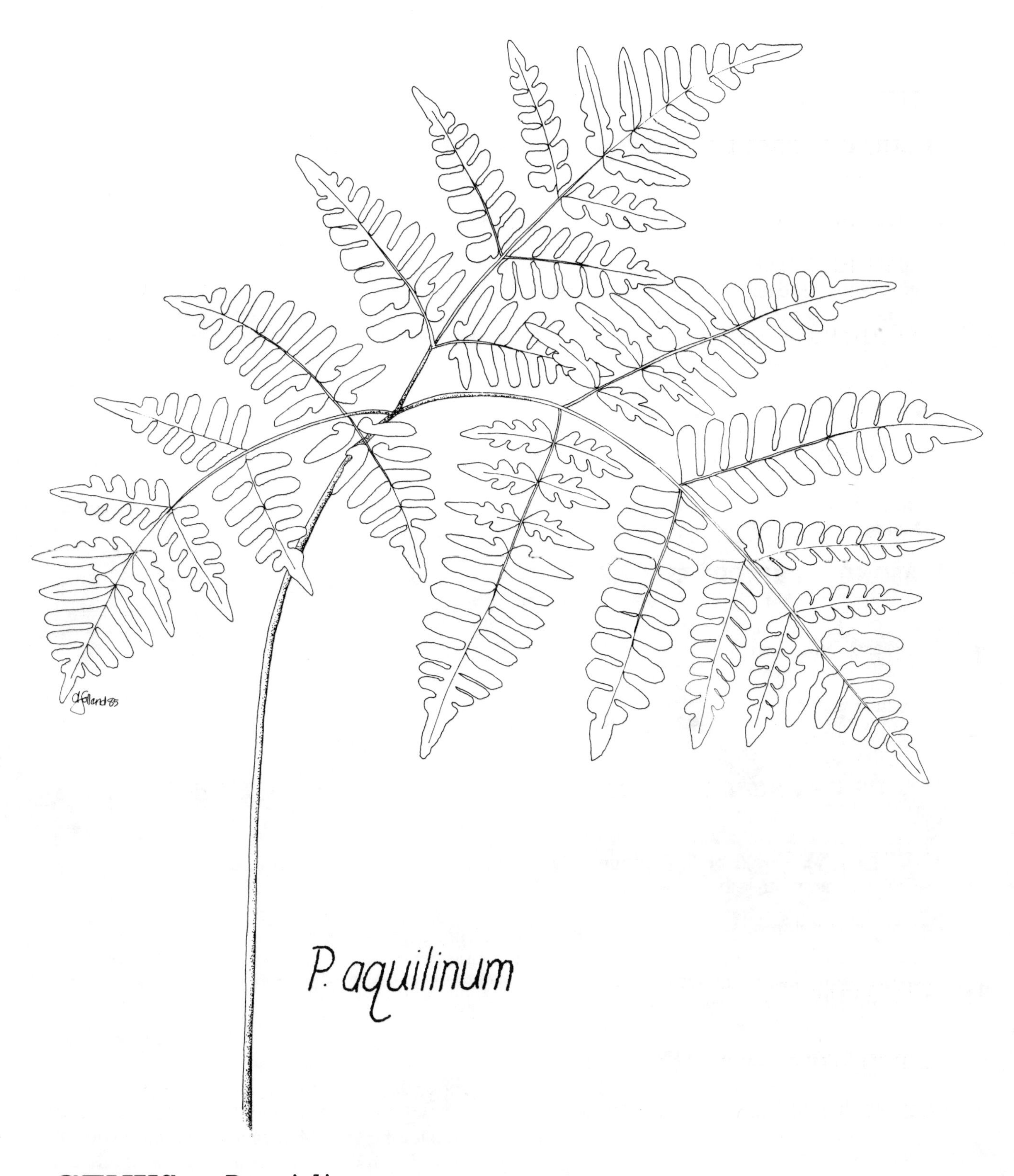


**GENUS:** *Pteridium*

*Pteridium aquilinum* (L.) Kuhn — Bracken fern; brake fern

**FAMILY:** Polypodiaceae — the Fern Family

This large family of ferns is delimited by the technical characters of the spore-bearing structures found either on the underside of fronds, or as separate, modified leaves.

**PHENOLOGY:** *Pteridium* produces spores in the summer.

**DISTRIBUTION:** Found in woods, thickets, clearings, and burned areas.

**PLANT CHARACTERISTICS:** Bracken grows 5-15 dm tall; **stem:** long, stout, erect, with a more spreading blade; **blade:** ternate and 2-3 times pinnately compound; **pinnae:** opposite or nearly so; **spores:** tetrahedral; **rhizome:** blackish, widely creeping with septate hairs, not scaly; **fronds:** leathery and deciduous.

**POISONOUS PARTS:** The entire plant is poisonous in a fresh or dried condition; dead fronds apparently are not harmful.

**SYMPTOMS:** Horses (and monogastric animals) show anorexia, bradycardia, and incoordination. The animal may crouch with feet apart and back and neck arched. With severe signs there is tachycardia; death occurs with clonic spasms. In ruminants (sheep and cattle) one can see a rough coat, listless attitude, and mucous nasal and oral discharges (possibly bloody) about one week before the serious symptoms occur. In acute cases, an elevated temperature appears. Also, there is anorexia and blood in excreta. In a prolonged illness, emaciation, hematuria, and rarely icterus can be observed. In young cattle, there is edematous swelling in the neck region with difficult breathing and death.

**Postmortem: gross lesions:** in monogastric animals: no significant gross lesions; enteritis with pericardial and epicardial hemorrhages; in ruminants: widespread petechiae and ecchymoses on serosal surfaces, mucosa, heart, muscles, and subcutaneous tissue. Abomasal ecchymoses may lead to ulceration; anemia and aplastic bone marrow are present; **histological lesions:** in monogastric animals: similar to those recorded for *Equisetum* poisoning; in ruminants: bladder lesions, ureters, or renal pelvis representing chronic severe hyperplasia and hemorrhagic inflammation that may lead to neoplasia. The transitional epithelium has a localized proliferation with metaplasia to mucinous columnar or stratified squamous types, or a combination of both. Hyperplastic epithelium develops neoplastic properties, transforming into a squamous cell or adeno carcinoma that is locally invasive and may spread to regional lymph nodes and lung. Hemorrhage in the urine may be a result of capillaries in the inflammatory lesion becoming hyperplastic and forming hemangiomas in the stroma or on the mucosal surface.

**POISONOUS PRINCIPLES:** The enzyme thiaminase is suspected in horses. In ruminants the agent of toxicosis is not known but causes hypoplasia or aplasia of hematopoietic tissue.

**CONFUSED TAXA:** This is the only fern that produces tall, large, coarse fronds from forking, extensively creeping rhizomes.

**TREATMENT:** Monogastric animals require thiamine treatment. Ruminants are treated with batyl alcohol, antibiotic, antiheparin, and antihistamine.

**SPECIES OF ANIMAL AFFECTED:** Horses, cattle, sheep, and possibly swine are susceptible.

**OF INTEREST:** It may take one to three months after ingestion for signs or symptoms to be manifest in thiamine-deficient animals. Six pounds per day for one month will poison a horse. Cattle fed hay with 50% bracken for 30-80 days will be poisoned; more is needed to poison sheep.

Other ferns known or suspected to be poisonous include sensitive fern (*Onoclea sensibilis* L.), which may produce nervous disorders (horses) and the male fern (*Dryopteris felix-mas* (L.) Schott.), which is suspected to contain thiaminase.

# Quercus

# GENUS: *Quercus*

*Quercus* spp. — Oaks

**FAMILY:** Fagaceae — the Beech Family

This economically important family contains the oaks, beeches, chestnut, and numerous other genera of **trees.** Characteristic features include **leaves:** alternate, simple, often toothed or cleft (lobed); **male flowers:** solitary or clustered.

**PHENOLOGY:** The oaks generally flower in mid-spring, the flowers appearing before the leaves.

**DISTRIBUTION:** Oaks are familiar trees found in a diversity of habitats from swamps to dry upland woods.

**PLANT CHARACTERISTICS:** Oak trees are very common and probably recognized by most readers; **leaves:** pinnately nerved; **fruit:** an acorn.

**POISONOUS PARTS:** Acorns and young shoots can cause severe poisoning, especially if eaten in quantity.

**SYMPTOMS:** Livestock display the following symptoms: anorexia, initial constipation (hard, dark fecal pellets) passing into diarrhea if the animal lives, gastroenteritis, thirst, and excessive urination.

**Postmortem: gross lesions:** lower half of digestive tract displays mucoid enteritis, becoming hemorrhagic; edema of mesenteric lymph nodes; subcutaneous edema and increased peritoneal and plural fluids; congested liver; gallbladder distended with viscid, brown bile; kidneys are enlarged, pale, uniformly covered with petechiae; **histological lesions:** brown-stained albumin in proximal convoluted tubules; necrotic epithelial lining cells mixed with proteinaceous substance such that the contents of the lumen form a dense homogeneous mass that is limited by the basement membrane and interstitial tissue.

**POISONOUS PRINCIPLES:** The toxins are unknown. Oaks contain large amounts of tannin (gallotannins), which has been implicated in poisonings. These substances are broken down into gallic acid and pyrogallol.

**CONFUSED TAXA:** Oaks are very common and not readily confused with other trees. One major forest type in Pennsylvania is the oak/hickory association. They are also planted for ornamental value.

**SPECIES OF ANIMALS AFFECTED:** Oaks are more a major problem on western rangeland than in Pennsylvania. Cattle, sheep, horses, and swine have been known to be poisoned. Human poisonings have not been reported.

**TREATMENT:** (11a)(b); (26)

**OF INTEREST:** Acorns and, to some extent buds, constitute a major source of wildlife food. Amerindians used acorns in their diet. White oak acorns were usually roasted and ground into a meal for use in "cakes." Books on edible wild plants often suggest eating acorns. Acorns from the white oak group apparently are palatable when cooked. The black oak/red oak group produces very bitter kernels. The white oak group can be differentiated from the red oaks by the presence of a scaly trunk, the tips and lobes of the leaves lacking bristly elongations, and the inside of the acorn smooth. Beech trees also are to be held suspect. Beech nuts, the fruit of *Fagus grandifolia* Ehrh., are reported edible in America; European authors, however, claim they have poisonous properties. They should be avoided, at least in quantity.

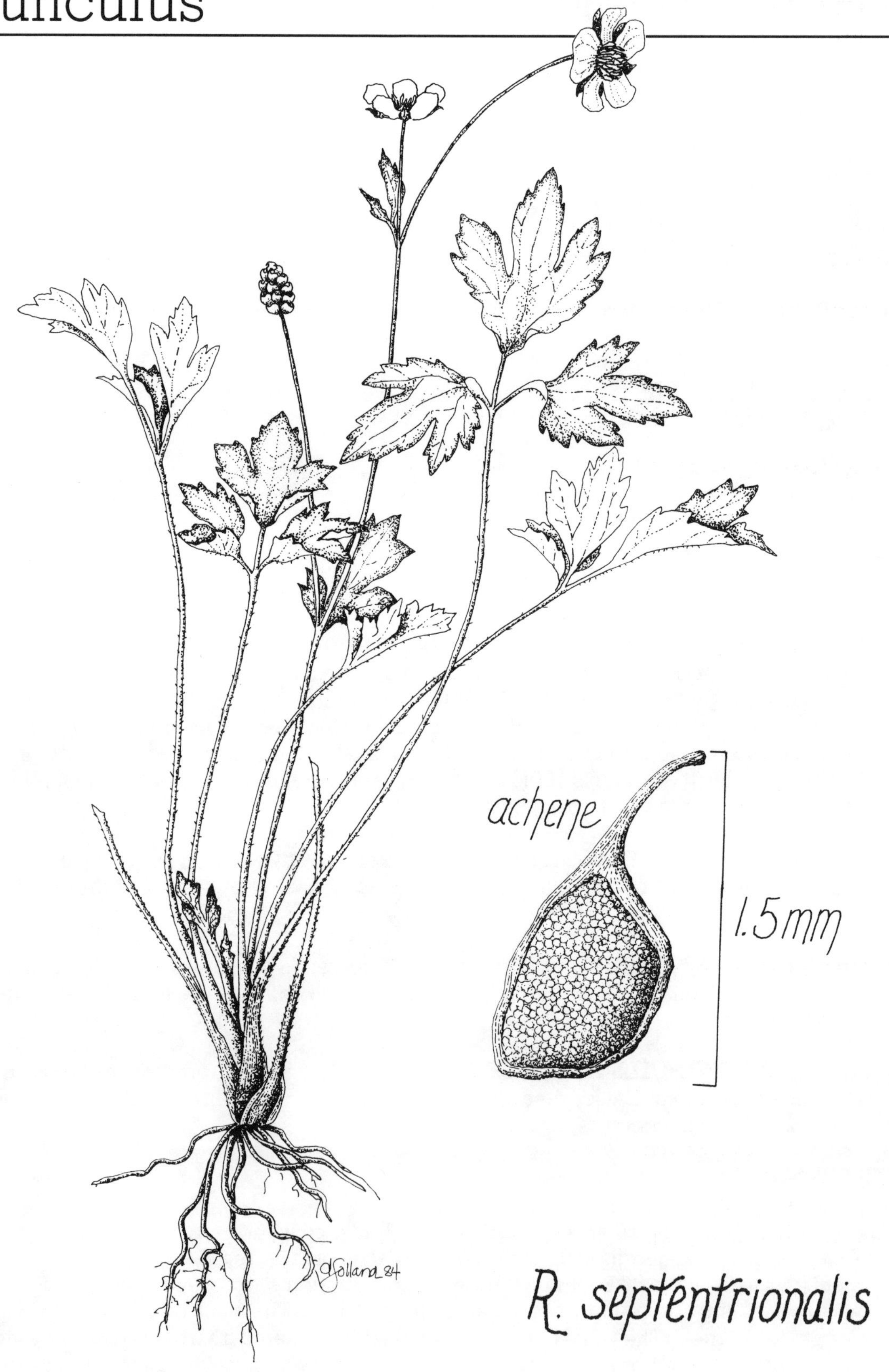
achene
1.5mm
d.gollana_84
R. septentrionalis

# GENUS: *Ranunculus*

*Ranunculus abortivus* L. — abortive buttercup; small-flowered crowfoot
*Ranunculus scleretus* L. — cursed buttercup
*Ranunculus septentrionalis* Poir. — Northern buttercup

---

**FAMILY:** Ranunculaceae — The Buttercup Family (see *Actaea*)

**PHENOLOGY:** The abortive buttercup flowers April and May, the cursed buttercup May through August, and the northern buttercup April through June.

**DISTRIBUTION:** The buttercups are common in Pennsylvania. The abortive buttercup inhabits moist or dry woods; the cursed buttercup, marshes, ditchbanks, and swampy meadows; and the northern buttercup, wet woods and meadows.

**PLANT CHARACTERISTICS:** *Ranunculus abortivus* has **petals:** yellow, 2-3 mm long, equal to or smaller than the sepals: **achenes:** with a very short beak, in a short, ovoid head on a villous receptacle. *R. scleratus* has **petals:** yellow, 2-3 mm, shorter than the sepals; **achenes:** nearly beakless, numerous in a short, cylindric head. *R. septentrionalis* has **petals:** yellow, 7-15 mm, about 2x longer than the sepals; **achenes:** with a straight beak, 1.8-3 mm long.

**POISONOUS PARTS:** Fresh leaves and the inflorescence are toxic. Dried material in hay reportedly is not poisonous. Toxicosis varies with amount ingested, stage of plant growth (most toxic at time of flowering), speed and degree of digestion or release of the toxin, growing conditions of the buttercup, and general health or susceptibility of the animal.

**SYMPTOMS:** Severe gastrointestinal irritation indicated by salivation, decreased appetite, colic, diarrhea, and slow pulse result from poisoning. Milk from affected cows may be bitter and/or reddish. Convulsions, sinking of eyes in their sockets, hematuria, and blindness are seen in severe cases. Horses, goats, and pigs show irritated tissues of the oral cavity. Convulsions may end in death. Buttercup toxicosis displays pulmonary congestion and ecchymotic hemorrhages on the pleural surfaces on **postmortem** examination.

**POISONOUS PRINCIPLES:** Toxicity is due to the unstable irritant oil protoanemonin. This oil is volatile and yellow due to lactone. *R. scleratus* contains the highest concentration of the oil.

**SPECIES OF ANIMAL AFFECTED:** All classes of livestock are susceptible.

**CONFUSED TAXA:** Numerous unrelated plants have 5 bright, yellow petals, although only *Ranunculus* petals are shiny and porcelainlike. Some members of the Rosaceae have flowers that might superficially resemble those of buttercups. However, the leaves of the rose family bear stipules, which are absent in *Ranunculus.*

**TREATMENT:** (11a)(b); (26); (5- at a rate of 2 mg subcutaneously, repeated as necessary); (27); (6)

**OF INTEREST:** Literature records a case of two heifers that were successfully treated for cursed buttercup poisoning. Upon returning to pasture, they selectively ate the *R. scleratus* despite ample presence of better forage. These cows apparently developed a desire for this plant after having eaten it as a part of their diet. However, buttercups usually are strongly distasteful to grazing animals and are eaten as a last resort after depletion of more desirable forage.

# GENUS: *Rheum*

*Rheum rhaponticum* L. — Rhubarb

**FAMILY:** Polygonaceae — the Smartweed Family

This family of plants contains at least 40 genera and more than 800 species, all with jointed **stems.** Other characters include **leaf stipules:** united into a tubular sheath called an ocrea; **sepals:** petaloid; **petals:** absent; **fruit:** an achene. The Polygonaceae are not known for their poisonous members but for useful ones such as buckwheat and various ornamental plants. Many elements in the family are weedy.

**PHENOLOGY:** Rhubarb flowers are borne on tall, hollow stalks in the summer.

**DISTRIBUTION:** *Rheum rhaponticum* is a cultivated plant that occasionally escapes from the garden.

**PLANT CHARACTERISTICS:** Rhubarb can be identified by **leaves:** large, basal, in clumps; ovate with cordate bases; **leaf blades:** up to 1.5 m long, margins wavy; **petioles:** as long as leaf blades, often red, stout; **sepals:** 6, greenish, whitish, or reddish; **stamens:** 6 (9); **fruit:** a 3-winged achene.

**POISONOUS PARTS:** The flat leaf blade is toxic.

**SYMPTOMS:** Human consumption of the rhubarb leaf results in gastroenteritis, cramps, nausea, vomiting, weakness, respiratory difficulties, irritation of the mouth and throat, poor clotting of the blood, internal hemorrhaging, coma, and death. In hogs the symptoms are staggering, salivation, convulsions, and death.

**POISONOUS PRINCIPLES:** Oxalic acid, uncharacterized soluble oxalates, and possibly other toxins are believed responsible for poisonings.

**CONFUSED TAXA:** Burdocks (see *Arctium*) are often confused with rhubarb. Burdock leaves are coarse and pubescent; the leaves of rhubarb are glabrous.

**SPECIES OF ANIMALS AFFECTED:** Rhubarb leaves are known to have caused the death of both humans and livestock.

**TREATMENT:** (11a)(b) with lime water, chalk, or calcium salts; (7); (26)

**OF INTEREST:** Several species of *Rheum* are grown for their bold foliage effects in landscaping. No data are available on the toxicity of these ornamental plants.

# GENUS: *Rhododendron*

*Rhododendron* spp. — Azalea, Rhododendron

**FAMILY:** Ericaceae - the Heath Family (see *Kalmia*)

**PHENOLOGY:** The azaleas and rhododendrons commonly flower in spring and early summer.

**DISTRIBUTION:** The genus *Rhododendron* can be divided into two nontechnical categories: cultivated and native plants. The cultivated plants are widely used around homes for their floral displays and, in some cases, evergreen foliage. The native plants are found in moist or wet woods, sometimes in dense colonies.

**PLANT CHARACTERISTICS:** Botanists recognize perhaps 800 species of *Rhododendron.* Included in this assemblage are azaleas, even though they usually are considered distinct by gardeners. The reader is generally familiar with cultivated rhododendrons and azaleas. The native, wild species do not differ greatly in appearance.

**POISONOUS PARTS:** The foliage of some species is toxic. All taxa are considered potentially poisonous.

**SYMPTOMS:** All poisonous members of the Ericaceae produce similar effects (see *Kalmia*). Symptoms can include salivation, tearing, nasal discharge, vomiting, convulsion and paralysis, and loss of appetite.

**POISONOUS PRINCIPLES:** The complex mixture andromedotoxin is suspected, as well as the hydroquinone glucoside arbutin.

**CONFUSED TAXA:** Rhododendrons and azaleas are familiar plants in Pennsylvania. No confusion is readily possible, with the exception of *Kalmia* (see *Kalmia*).

**SPECIES OF ANIMALS AFFECTED:** Several species of *Rhododendron* are known to cause loss of livestock.

**TREATMENT:** (11a)(b); (1); (5); (12)

# Rhus radicans

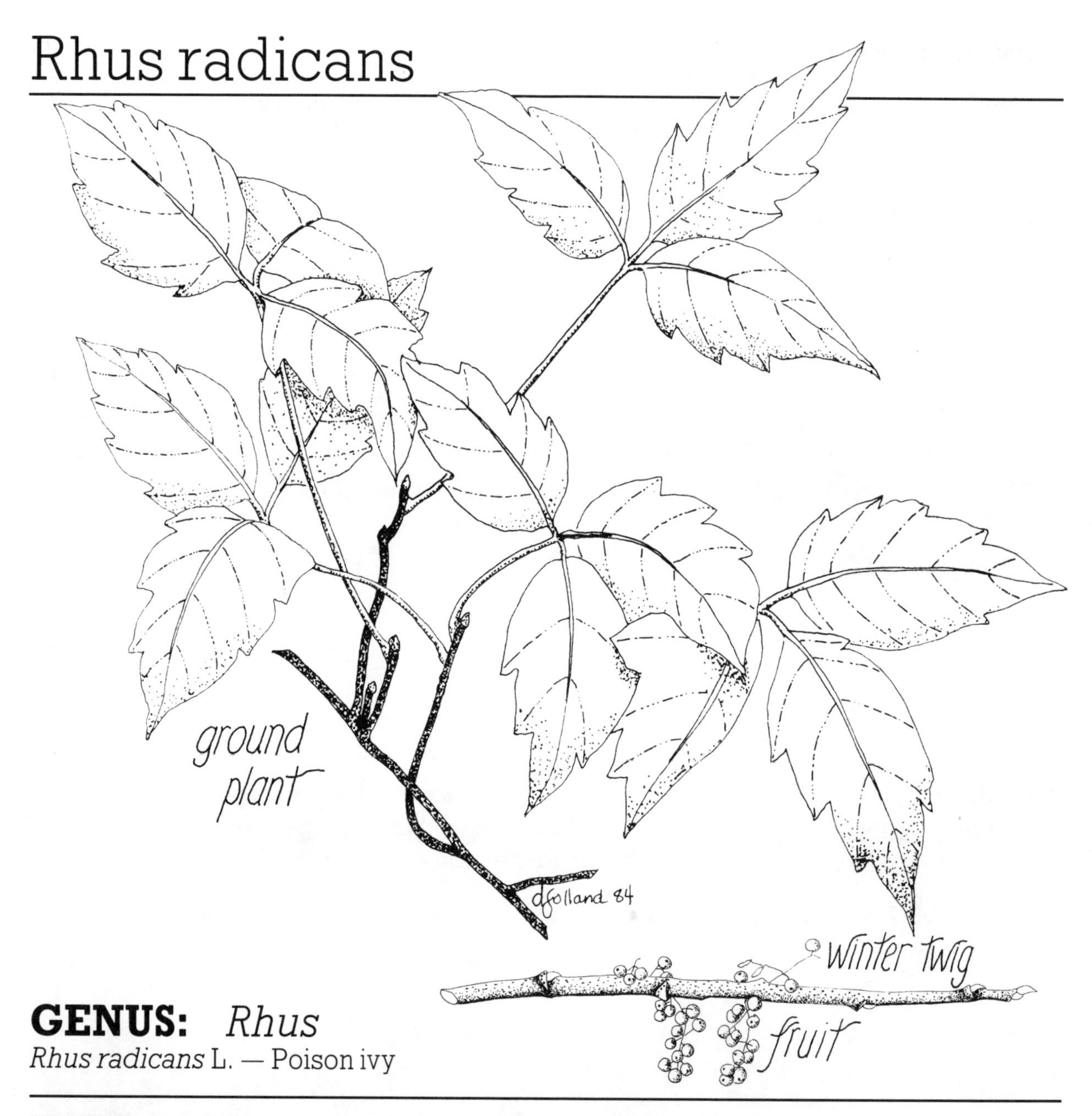


## GENUS: *Rhus*

*Rhus radicans* L. — Poison ivy

**FAMILY:** Anacardiaceae — the Cashew Family

Many readers will be surprised to learn that the edible cashew nut belongs to the same family of plants as poison ivy and poison sumac. This predominantly tropical family has **leaves:** alternate, compound; **flowers:** 5-merous, polypetalous, regular, with an annular disc between the 5 **stamens** and ovary; **ovary:** 1-celled, containing 1 ovule; **fruit:** a drupe.

**PHENOLOGY:** Poison ivy produces inconspicuous greenish flowers from May through July.

**DISTRIBUTION:** *Rhus radicans* is commonly found in disturbed habitats, flood plains, cultivated fields, cemeteries, waste places, along woodland paths, margins of woodlots, fencerows, road-banks, along streams, and in urban situations around buildings and yards.

**PLANT CHARACTERISTICS:** The species has complex and variable forms. Some are woody vines that produce aerial roots and grow by straggling and climbing over other vegetation. Ground-forms usually spread by rhizomes and develop dense colonies with a few leaves crowded near the summit. Regardless of growth habit, poison ivy always has three leaflets per

**leaf,** with **leaflets:** ovate to subrotund, varying to rhombic or elliptic, terminally acute to acuminate, basally cuneate; entire to irregularly serrate or crenate; glabrous or thinly pubescent, **petiolule** of the terminal leaflet longer than those of the lateral leaflets; **panicles:** axillary, 1 dm long, bearing greenish-yellow flowers that mature into grayish white fruits, 5-6 mm; **fruits:** mature August through November, conspicuous all winter; birds eat the ripe seeds with impunity.

**POISONOUS PARTS:** All parts of poison ivy, with the possible exception of the pollen, contain toxins that cause dermatitis. It has been suggested that extremely sensitive persons might contract poison from wind-blown pollen in spring when the plant is flowering.

**SYMPTOMS:** Dermatitis ranging from minor reddened and itching skin to major swelling, blisters, and weeping wounds can result from contact. Ingestion of leaves can cause irritation of the mucosa and digestive tract; gastritis and death may result. Animals probably are not as susceptible as humans to contact dermatitis due to hair and fur. Ingestion of leaves or other plant parts by livestock could be dangerous and result in death.

**POISONOUS PRINCIPLES:** The toxin 3-n-pentadecylcatechol has been isolated from *Rhus radicans.*

**CONFUSED TAXA:** The plant most commonly confused with poison ivy is Virginia creeper, *Parthenocissus quinquefolia, which produces berries that are poisonous upon ingestion (see Parthenocussus*). A reliable feature useful in differentiating the two plants is the compound leaf. Virginia creeper has 5 leaflets, whereas poison ivy has 3 leaflets per leaf. Ash tree species in the genus *Fraxinus,* and boxelder (*Acer Negundo* L.), can superficially resemble poison ivy, especially as seedlings; however, the former two have opposite leaves, whereas poison ivy has alternate leaves.

**SPECIES OF ANIMALS AFFECTED:** Contact dermatitis is commonly seen in humans. Fifty percent of the population is allergic in some degree to poison ivy. Consumption of this plant will affect humans, and probably livestock as well.

**TREATMENT:** For dermatitis: (4); (23); in severe cases steroid injections can reduce the reaction.

**OF INTEREST:** The following concerning poison ivy are true. They are listed to dispel popular myths.

- Poison oak is not found in Pennsylvania.
- Poison ivy plants contain the skin irritant all year.
- The plant must be bruised or broken for the toxin to exude from the plant; there is no substantial evidence for the plant otherwise exuding the poisonous principles.
- The toxin is not volatile nor air borne except when carried in droplet form on smoke, dust, or combusted particles generated by burning the plant.
- Weeping wounds and blisters do not spread poison ivy over the body.
- Towels or clothing contaminated by the serum from weeping wounds will not spread the itching. New blisters are the result of delayed response at the site of infection, renewed contact with the plant, or recontact with irritant-contaminated articles.
- The irritating chemical can be spread from contaminated articles, clothing, pets, garden tools, etc.
- After contact with the irritant, symptoms may appear within hours or up to a week later.
- Poison ivy plants can be eliminated by herbicide application. These compounds are generally nonselective and may kill surrounding vegetation if applied improperly.

For a list of currently registered herbicides contact the Pennsylvania Department of Agriculture, the County Agricultural Extension Office, a local lawn and garden center, or agricultural farm products supplier.

# Rhus vernix

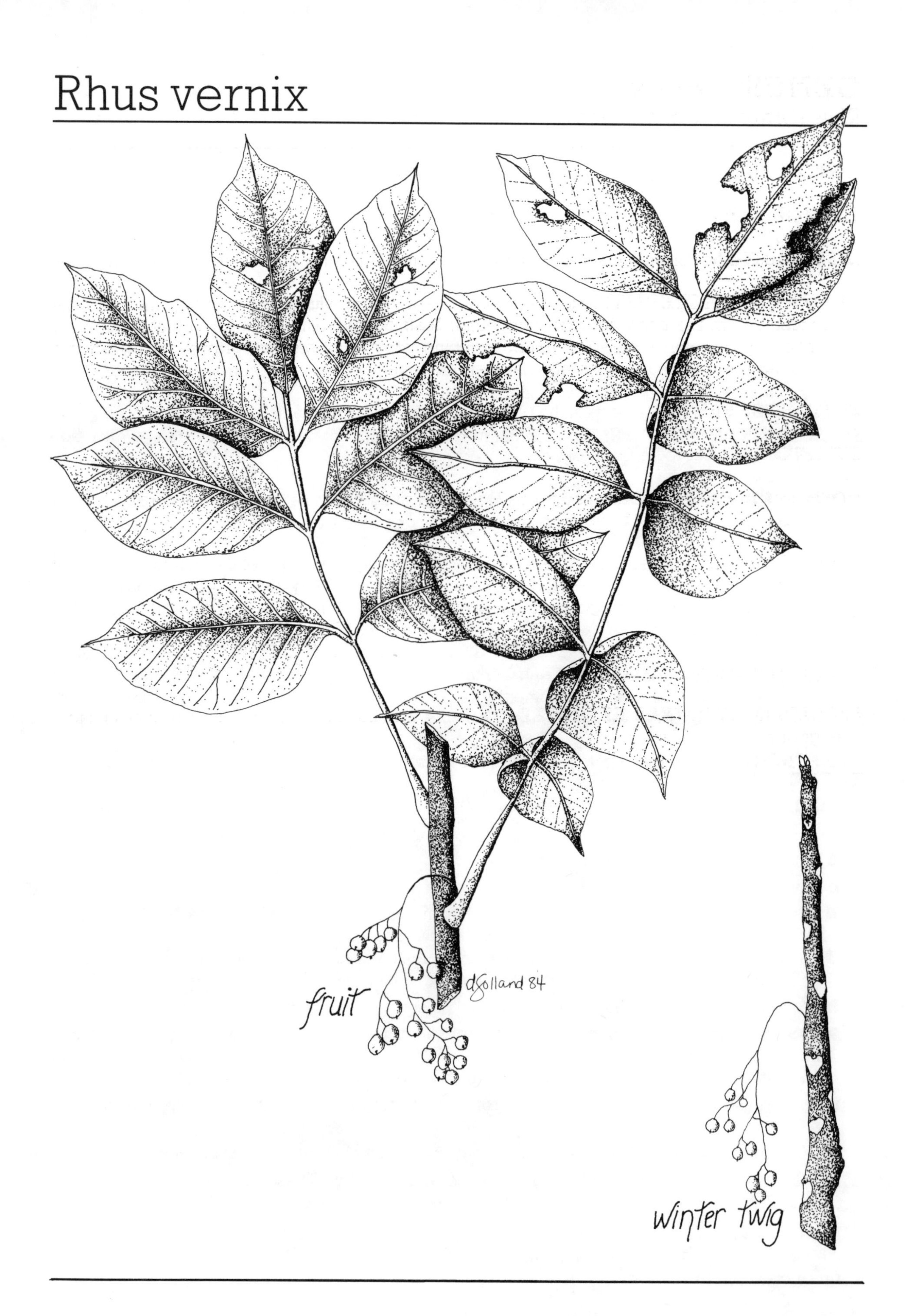

# GENUS: *Rhus*

*Rhus Vernix* — Poison sumac

**FAMILY:** Anacardiaceae — the Cashew Family (see *Rhus radicans*)

**PHENOLOGY:** Poison sumac flowers May through July.

**DISTRIBUTION:** Poison sumac is found in bogs, swamps, marshes, and shaded wooded wetlands.

**PLANT CHARACTERISTICS:** Poison sumac is a swamp **shrub** growing to 5 meters, often branched from the base; **leaves:** compound; **leaflets:** odd-pinnate, 7-13 per leaf, oblong to elliptic, 4-8 cm, entire, glabrous; **fruit:** drooping panicles of berries, grayish white, 4-5 mm; produced August through November, evident all winter.

**POISONOUS PARTS:** All parts of the plant contain the contact irritant.

**SYMPTOMS:** Refer to *Rhus radicans.*

**POISONOUS PRINCIPLES:** The toxins for poison sumac have not been characterized but are probably similar to those found in poison ivy.

**CONFUSED TAXA:** There are several "sumacs" in Pennsylvania. All of the nonpoisonous ones have erect, not pendulous fruits, and are found in drier soil. They also have toothed or serrate leaflets unlike the entire margin in poison sumac. The Tree-of-heaven (*Ailanthus altissima* (Mill.) Swingle) is a rapidly growing, weedy tree common in cities. The leaflets of *Ailanthus* have one or more coarse, basal teeth, each with a large gland beneath. In addition, tree-of-heaven produces winged fruits with a central seed.

**SPECIES OF ANIMALS AFFECTED:** Probably only humans will encounter poison sumac in bogs or swamps.

**TREATMENT:** Refer to *Rhus radicans.*

# Ricinus

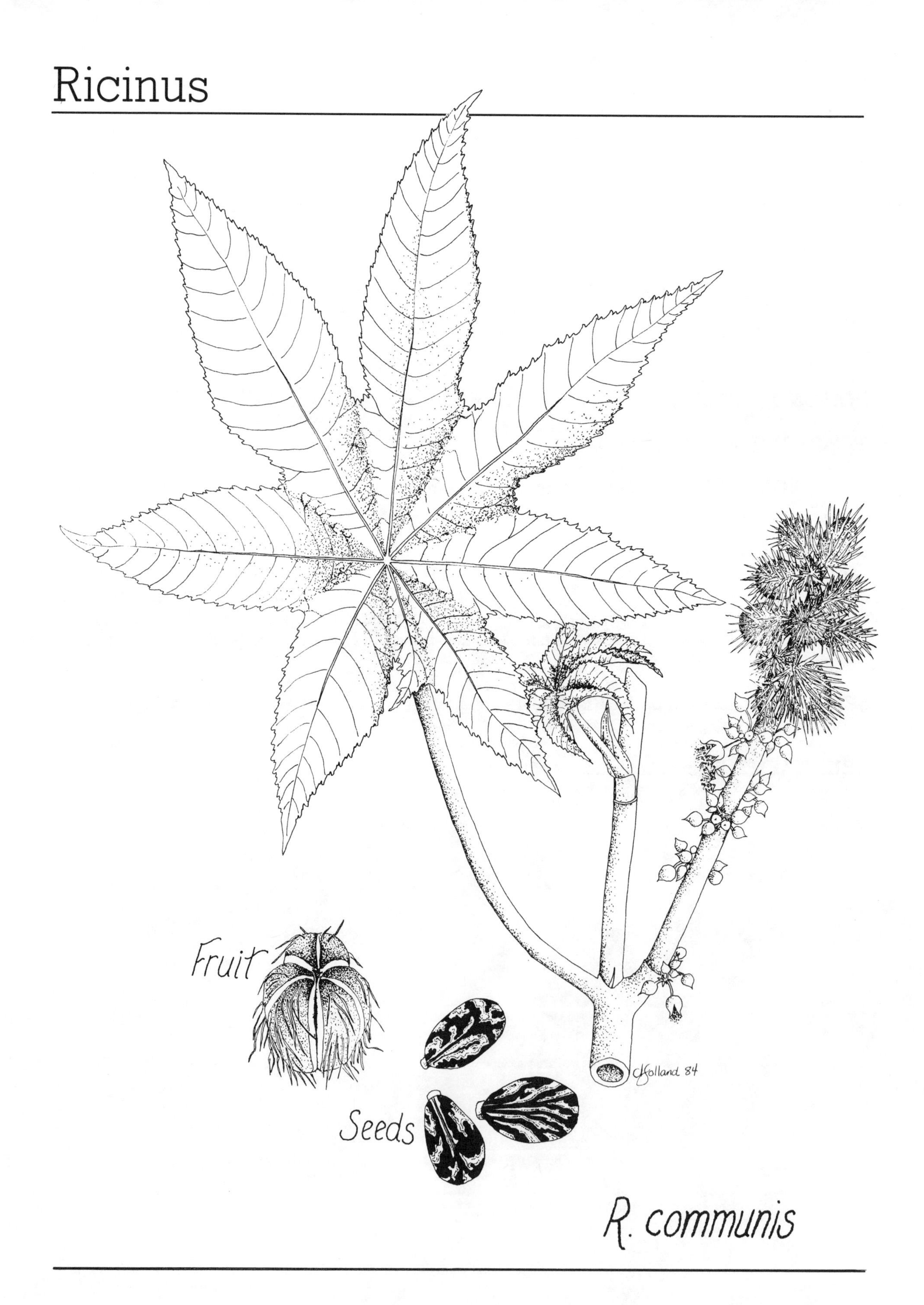

# GENUS: *Ricinus*

*Ricinus communis* L. — Castor bean; Palma Christi

---

**FAMILY:** Euphorbiaceae — the Spurge Family (see *Codiaeum*)

**PHENOLOGY:** Flowering is dependent on the time of year that the seeds are sown. Seeds may be sown in the place where the plants are to be grown or in pots and transplanted in mid-May. Flowering occurs in mid-summer.

**DISTRIBUTION:** This plant, introduced from the African tropics, is sometimes sown in gardens in the state for its rapid growth and bold, striking colors.

**PLANT CHARACTERISTICS:** Castor bean is treated as an annual. It can grow to 5 m tall. The **flowers** are monoecious and without petals; **stamens:** very numerous, filaments much-branched; **ovary:** 3-celled; **style:** 3, each bifid and plumose, united at the base; **fruit:** a large, 3-lobed capsule covered with soft prickles; **seeds:** 1 cm long, mottled or streaked with white, red, or brown; **leaves:** alternate, large, simple, peltate, palmately veined, long petiolate, palmately 6- to 11-lobed; 1-4 dm wide.

**POISONOUS PARTS:** Seeds, and to a lesser extent foliage, are toxic; 1-3 seeds may be fatal to a child, 2-4, to an adult.

**SYMPTOMS:** There is often a lag time from initial ingestion until symptoms appear. Poisoning is indicated by gastrointestinal distress, burning mouth and throat, anorexia, nausea and vomiting, cramps, cessation of rumination, dullness, diarrhea, weakness, thirst, prostration, dullness of vision, convulsion, muscle spasm, uremia, and death. The digestive tract displays inflammation and punctiform hemorrhage of the mucosal lining. Organ damage includes fluid-filled lungs, and edematous and swollen liver and kidneys. In horses trembling, sweating, and incoordination may precede other symptoms, accompanied by unusually vigorous heart contractions and weak, rapid pulse. Cattle may display blood-stained diarrhea; pigs vomit profusely. In poultry egg production ends, molting commences, wattles and combs discolor, and the birds appear depressed.

**Postmortem: gross lesions:** mesenteric lymph nodes are edematous; **histological lesions:** necrosis of epithelium of affected gastrointestines; hydropic degeneration, fatty change, and necrosis of hepatocytes; renal epithelium experiences fatty degeneration and necrosis; marked destruction of lymphocytes in lymphoid organs; brain necrosis; in horses edema (pulmonary, bronchial, mesenteric, and hepatic lymph node).

**POISONOUS PRINCIPLES:** The highly toxic glycoprotein ricin is responsible for poisoning. This phytotoxin, a composite of various amino acids, consists of a neutral alpha-chain capable of inhibiting protein synthesis and an acidic beta-chain, which functions as a carrier and moiety that binds the toxin to cell surface. Phytotoxins may act as antigens eliciting an antibody response.

**CONFUSED TAXA:** No other ornamental plants have the characteristics described above; castor bean is readily distinguishable.

**SPECIES OF ANIMALS AFFECTED:** Humans are susceptible to this highly active poison. Seed amounts necessary for poisoning depend on the age of victim and amount of seed masticated since chewing enhances liberation of the toxin.

**TREATMENT:** Immediate (11a)(b); (26); (13 by administering 5-15 gm sodium bicarbonate daily)

**OF INTEREST:** The seeds of castor bean yield a familiar oil used extensively in industry for the manufacture of soap, varnishes, and paints. This plant also has medicinal qualities.

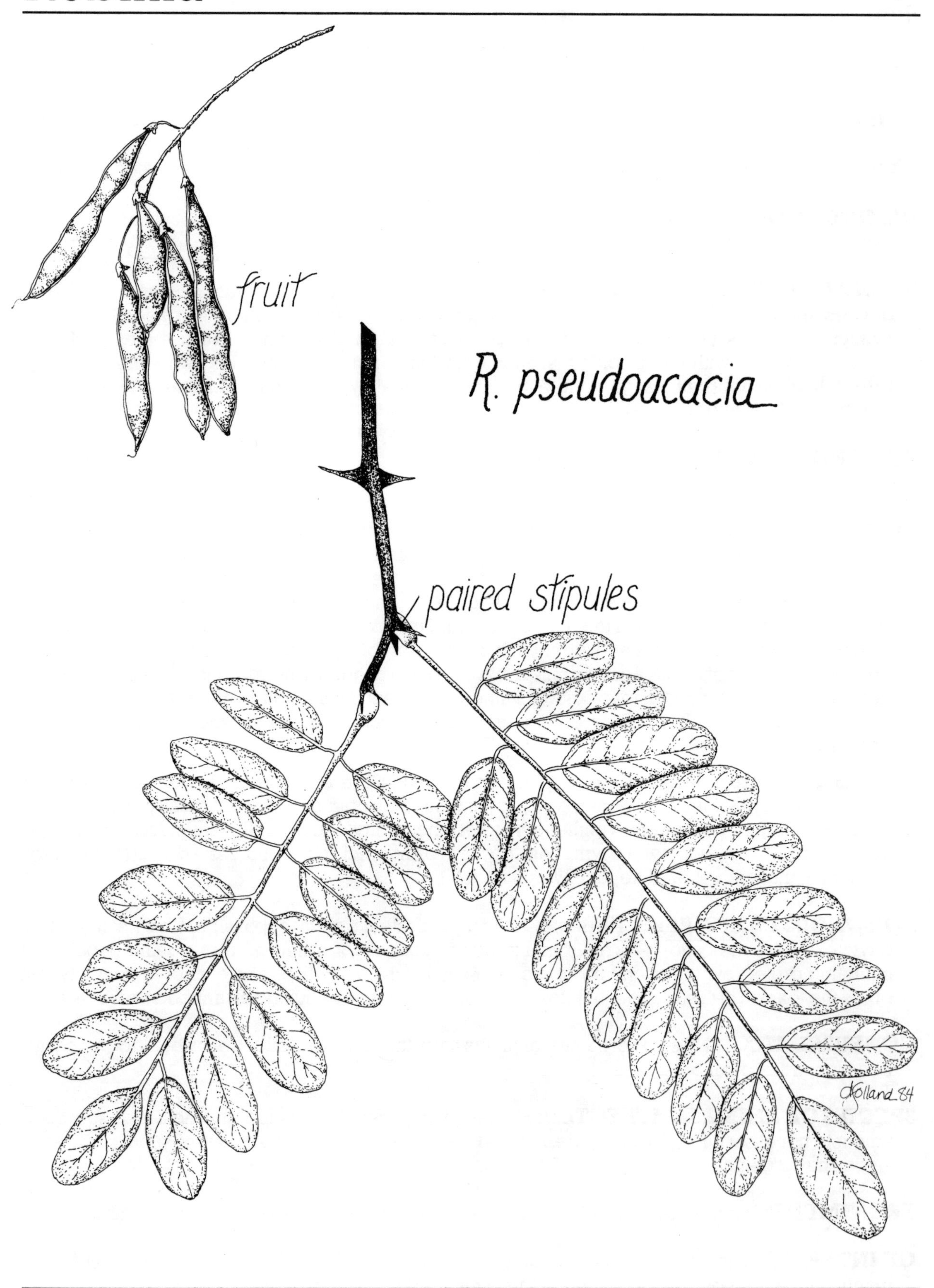
fruit
R. pseudoacacia
paired stipules

# GENUS: *Robinia*

*Robinia pseudoacacia* L. — Black locust

**FAMILY:** Fabaceae (Leguminosae) — the Bean Family (see *Crotalaria*)

**PHENOLOGY:** Long, fragrant, white, grapelike clusters of flowers are produced in May and June.

**DISTRIBUTION:** *Robinia pseudoacacia* is a native tree inhabiting woods, thickets, and fencerows.

**PLANT CHARACTERISTICS:** Black locust is an open **tree** that can reach a height of 25 m; **stipules:** modified into two opposing spines 1 cm long flanking the base of the petiole; **leaves:** odd-pinnately compound; **leaflets:** up to 9 pairs, elliptic to ovate-obtuse, 2-4 cm; **inflorescence:** a raceme, dense, drooping, many-flowered, 20 cm long; **flowers:** white, fragrant, 2-2.5 cm; **fruit:** 5-10 cm long, reddish brown, glabrous, remaining in clusters on the tree over winter and becoming black.

**POISONOUS PARTS:** Toxins are produced by the plant and accumulate in the leaves, seeds, and inner bark. In controlled experiments on horses an aqueous extract of bark (0.1% of body weight) and powdered bark (0.04% of the body weight) were found toxic. Poisonous principles are about one-tenth as toxic to cattle.

**SYMPTOMS:** Toxic reactions include weakness, depression, anorexia, vomiting, diarrhea (blood may be present), nausea, dilated pupils, coldness of extremities, and weak and irregular pulse. Lesions include irritation and edema of the digestive mucosa and severe gastroenteritis. There may be venous congestion. Also, a yellowish pigmentation of the membranes, similar to icterus, may be present. Fatalities are rare.

**POISONOUS PRINCIPLES:** Two compounds are suspected to be involved in toxicity, a heat-labile phytotoxin, robin, and the glycoside robitin. Additional isolated compounds include acetin and robinetin; it is uncertain what role these substances play in toxicosis.

**CONFUSED TAXA:** Two other species of *Robinia* are found in Pennsylvania: *R. hispida* L. and *R. viscosa* Vent. Both species have pink flowers and seed pods with stiff spreading hairs. The honey locust, *Gleditsia triacanthos* L., a tall tree with branched thorns, has more narrow (oblong-lanceolate) leaflets that are obscurely crenate. The leaves of this plant are even-pinnate or bipinnate and the flowers unisexual. Honey locust seed pulp is considered to be a pleasant tasting, sweet nibble but should not be confused with the poisonous pulp of Kentucky coffeetree (see *Gymnocladus*).

**SPECIES OF ANIMALS AFFECTED:** Humans and a wide variety of livestock have shown symptoms. These include horses, cattle, sheep, and poultry.

**TREATMENT:** (11a)(b); (26); (13)

**OF INTEREST:** This plant is listed in some texts as an emergency-food: inner bark, flowers fried or infused in water for a beverage. Extreme caution should be exercised in this regard. Other species may be toxic. Clammy locust (rose-acacia), *R. viscosa* Vent., a native in mountain woods, and bristly locust (also called rose-acacia), *R. hispida* L., a roadside and spoil-bank taxon, may be poisonous.

# Sanguinaria

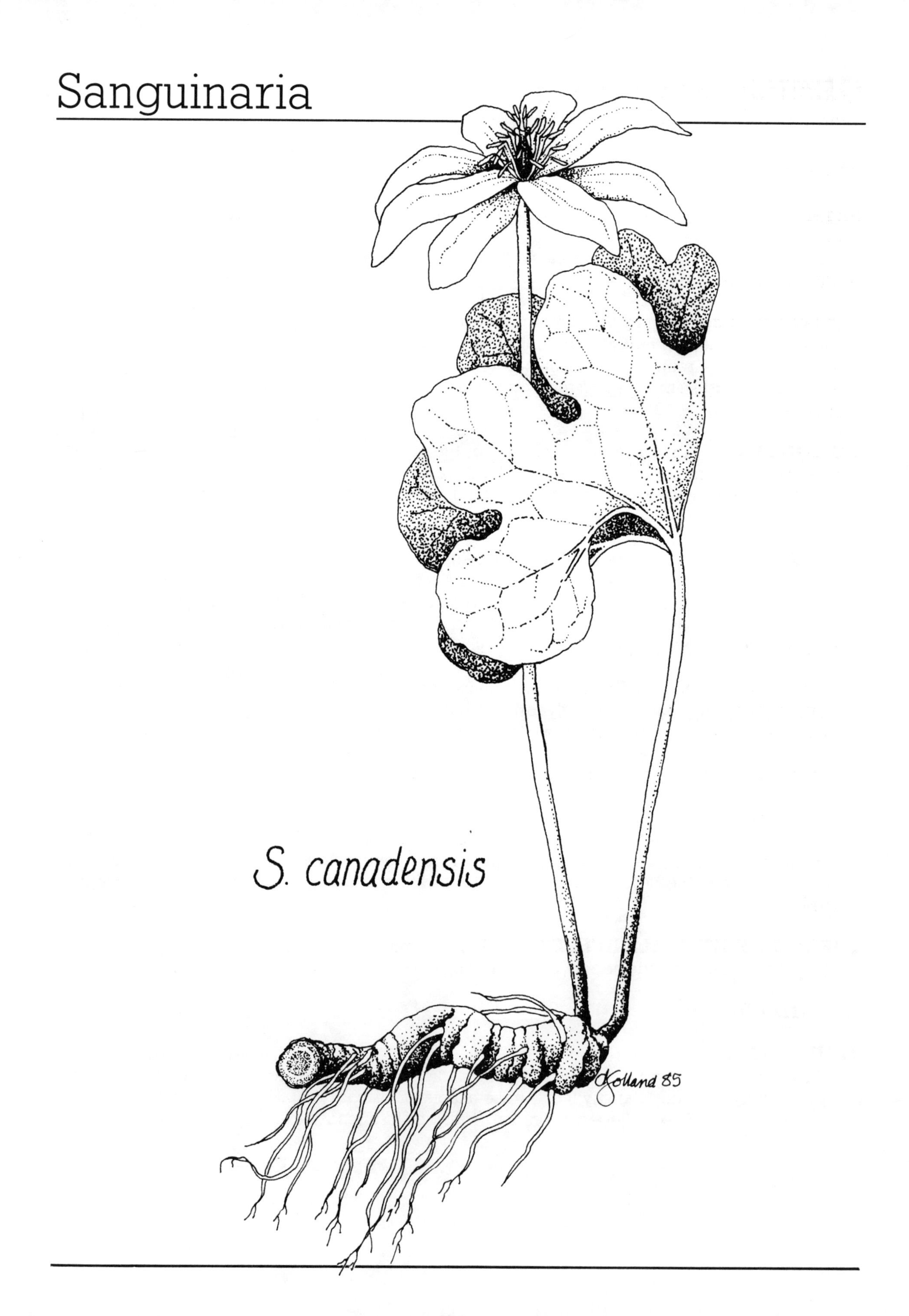

# GENUS: *Sanguinaria*

*Sanguinaria canadensis* L. — Blood root; red puccoon

---

**FAMILY:** Papaveraceae — the Poppy Family (see *Chelidonium*)

**PHENOLOGY:** Bloodroot is an early spring plant, with flowers appearing in April before the leaves.

**DISTRIBUTION:** *Sanguinaria canadensis* is distributed in rich woods.

**PLANT CHARACTERISTICS:** This perennial plant grows from a stout, knotted **rhizome** that sends up a large white flower on a scape, with a single leaf: orbicular in outline, up to 2 dm wide at maturity in late season, 3-9 lobed; **scape:** at flowering 5-15 cm; **flowers:** 2-5 cm wide; **sepals:** 2, falling early; **petals:** 8-16, 4 usually longer than the others and the flowers quadrangular in outline; **stamens:** numerous; **ovary:** narrow, style terminated by a capitate, 2-lobed stigma; **fruit:** a fusiform capsule, 3-5 cm long, crowned by the persistent **style;** root, scape, petiole, and leaves bleed a red-orange **latex** when bruised.

**POISONOUS PARTS:** The entire plant contains alkaloid-laden red latex.

**SYMPTOMS:** The papaveraceous alkaloids can cause dropsy and glaucoma in humans. Loss of human life and livestock have been reported after consumption of plants containing these alkaloids. Symptoms include, vomiting, diarrhea, fainting, shock, and coma. Under natural conditions, no cases of bloodroot poisoning are known.

**POISONOUS PRINCIPLES:** Physiologically active latex constituents include sanguinarine (pseudochelerythrine), chelerythrine, protopine, homochelidonine, and resin.

**CONFUSED TAXA:** No spring-flowering plants from rich woods can readily be confused with bloodroot.

**SPECIES OF ANIMALS AFFECTED:** Humans and livestock are poisoned by poppy alkaloids.

**TREATMENT:** (1 1a)(b); (26)

**OF INTEREST:** The colored latex was used by Amerindians for painting skin and arrowshafts. The alkaloid sanguinarine from this plant is used in research to induce glaucoma in laboratory animals.

# GENUS: *Saponaria*

*Saponaria officinalis* L. — Soapwort; bouncing Bet

---

**FAMILY:** Caryophyllaceae — the Pink Family (see *Agrostemma*)

**PHENOLOGY:** Bouncing Bet flowers from July through September.

**DISTRIBUTION:** Formerly cultivated, bouncing Bet is now a weed of roadsides, waste places, and along railroads.

**PLANT CHARACTERISTICS:** Soapwort is a perennial, often colonizing larger areas by rhizomes; **plants** grow 4-8 dm and have smooth jointed **stems** with **leaves:** opposite, 7-10 cm long x 2-4 cm wide, without petioles, palmately veined (sometimes appearing parallel); **flowers:** congested, conspicuous, in large terminal clusters; **calyx:** 1.5-2.5 cm, the 5 lobes triangular with drawn-out tips, the tube often becoming deeply bilobed; **corolla:** 5 white-or pinkish-appendaged petals; **stamens:** 10, exsert; **styles:** 2; **capsules:** dehiscent by 4 (or 6) teeth; **seeds:** plump, kidney-shaped, small.

**POISONOUS PARTS:** The plants, especially the seeds, are poisonous. Laboratory feeding experiments have produced toxicity and death in rabbits. Sheep fed bouncing Bet in an amount equivalent to 3% (dry weight basis) of their weight died within four hours.

**SYMPTOMS:** Similar to those for *Agrostemma.*

**POISONOUS PRINCIPLES:** A saponin, sapogenin, is similar or equivalent to githagenin found in corncockle.

**CONFUSED TAXA:** Many species in this family superficially resemble *Saponaria.* The number of styles is helpful in distinguishing several similar genera. *Saponaria* has 2 styles; *Silene* (catchfly, campion) has 3 (or 4) styles; *Lychnis* (white campion) and *Agrostemma* (corncockle) have 5 styles.

Some authors separate the annual cow-herb, from the perennial bouncing Bet, calling the annual species *Vaccaria segetalis* (Neck.) Garcke. Others place the annual plants in *Saponaria,* calling them *S. vaccaria* L. Regardless of nomenclature, the bouncing Bet has a 20-nerved tubular calyx and appendaged petals, whereas cow-herb has a strongly wing-angled, ovoid calyx and petals lacking appendages. This troublesome weed of grain crops also is considered poisonous.

**SPECIES OF ANIMALS AFFECTED:** Both species (bouncing Bet and cow-herb) are unpalatable and generally avoided by animals. No clear cases of poisoning have been recorded in North American literature.

**TREATMENT:** (11a)(b); (26)

**OF INTEREST:** In North American folklore, decoctions of *Saponaria officinalis* were used as poultices to remove discoloration around black or bruised eyes. This plant has been used in European countries as a soap substitute.

# GENUS: *Solanum*

*Solanum carolinense* L. — Horsenettle
*Solanum nigrum* L. — Black nightshade; deadly nightshade; common nightshade; garden nightshade

---

**FAMILY:** Solanaceae — the Nightshade Family (see *Datura*)

**PHENOLOGY:** *Solanum carolinense* and *S. nigrum* flower May through October.

**DISTRIBUTION:** Both are found in disturbed soil, woods, meadows and pastures, and cultivated fields; *S. carolinense* is also found in barren fields and wasteland.

**PLANT CHARACTERISTICS:** Horsenettle can be recognized by its prickly, stellately pubescent appearance. **Plants** are rhizomatous, to 1 m with **leaves:** 7-12 cm, half as wide, with 2-5 large teeth or shallow lobes, prickly on veins beneath, elliptic to ovate; **inflorescence:** several-flowered, elongating at maturity to a simple racemiform cluster; **flowers:** 2 cm wide; corolla: violet to white; **anthers:** equal; **fruit:** yellow, 1-1.5 cm, subtended but not enclosed by the unarmed calyx.

Black nightshade is a branching annual, 1.5-6 dm, glabrous or somewhat strigose above; **leaves:** irregularly blunt-toothed or subentire, ovoid to deltoid, 2-8 cm; **flower:** white **corolla,** 5-10 mm; **fruit:** black, globose, 8 mm, mature **calyx** 2-3 mm, lobes often unequal.

**POISONOUS PARTS:** The berries and vegetation are poisonous. The toxicity is not lost in drying and may be toxic in hay.

**SYMPTOMS:** In sheep, severe intestinal lesions develop as a result of horsenettle toxicosis. There may be inflammation of the mouth and esophagus in calves. Nervous symptoms may include apathy, drowsiness, salivation, dyspnea, trembling, progressive weakness or paralysis, prostration, and even unconsciousness. In humans, loss of senses sometimes occurs. Gastrointestinal effects may include anorexia, nausea, colic, vomiting, and constipation or diarrhea (possibly with blood). Poisoning is not always fatal (fatalities are due to paralysis).

**Postmortem: gross lesions:** kidneys surrounded by blood-tinged serum and edema; toxicosis of longer duration produces blood clots; **histological lesions:** pale kidneys with toxic tubular necrosis with casts and proteinaceous precipitate in the lumen; focal hemorrhages and edema associated with the toxic nephrosis; digestive tract lesions include acute catarrhal or hemmorrhagic gastritis and enteritis with ulcers that may extend to or throughout the muscularis propria.

Ingestion of black nightshade may cause nervous symptoms including apathy, drowsiness, salivation, dyspnea, trembling, progressive weakness or paralysis, prostration, and loss of consciousness. In humans, stupefaction and loss of senses develop. Gastrointestinal irritation may include anorexia, nausea and vomiting, cholic, and constipation or diarrhea (diarrhea may contain blood). Poisoning does not always end in death. Toxicosis climaxes in a number of hours, or in 1 to 2 days. Death is the result of paralysis. Chronic poisoning may occur and may include ascites as a symptom.

**POISONOUS PRINCIPLES:** Solanine, a saponic glycoalkaloid that breaks down into a sugar (solanose) and an alkamine (solanidine), is responsible for poisoning. The alkamines are steroidal. Concentration of solanine may increase 10 times with maturity.

**CONFUSED TAXA:** There are approximately 10 species of *Solanum* encountered in Pennsylvania. The taxa are separated by technical characters such as pubescence, leaf and corolla shape, and calyx structure. The genus *Physalis* (ground cherry) is sometimes confused

*(Continued)*

# Solanum

with *Solanum*. *Physalis* has longitudinally dehiscent anthers and a spineless mature calyx. *Solanum* has anthers dehiscing by terminal pores and often a spiny calyx. The unripe fruits of *Physalis* are poisonous.

**SPECIES OF ANIMALS AFFECTED:** All species of livestock, deer, and humans are susceptible.

**TREATMENT:** 11(a) (b); (26); (17)

**OF INTEREST:** The dried berries of horsenettle, which cling to the plant over the winter, killed cattle in March. The berries may be sought in preference to other food. Horsenettle is believed to have caused the death of a 6-year-old boy in Delaware County, Pennsylvania, in 1963. *Solanum carolinense* seeds are Restricted Noxious Weed Seeds and must be listed on the tag or label on agricultural seeds sold in Pennsylvania.

Nightshade berries have been cooked and used for plum puddings and in preserves, jams, or pies with no ill effects. Boiling apparently destroys the toxic principle. When three kilograms of green plant were experimentally fed to a horse, no serious symptoms were observed.

The compound solanine has been used as an agricultural insecticide. The $LD_{50}$ i.p. in mice is 42 mg/kg. The cultivated house plant *Solanum pseudocapsicum* L. (Jerusalem cherry, Natal cherry) is a cardiac depressant. The leaves contain the cardioactive substance solanocapsine, while berries contain the glycoalkaloid solanine and related substances. This ornamental, potted plant will affect house pets and children if eaten. If grown too close to the soil surface, ordinary potatoes (*S. tuberosum* L.) will develop a green skin from exposure to the sun. This green skin, as well as young sprouts, can contain alkaloids that cause human and livestock toxicosis and fatalities. Brown-skinned, unsprouted potato tubers contain 0.009% solanine. Toxicosis is associated with concentrations of 0.04%. Green potatoes should not be used in food preparation, or the green tissue should be removed before the tuber is used. Symptoms of poisoning are those given above for the alkaloid solanine.

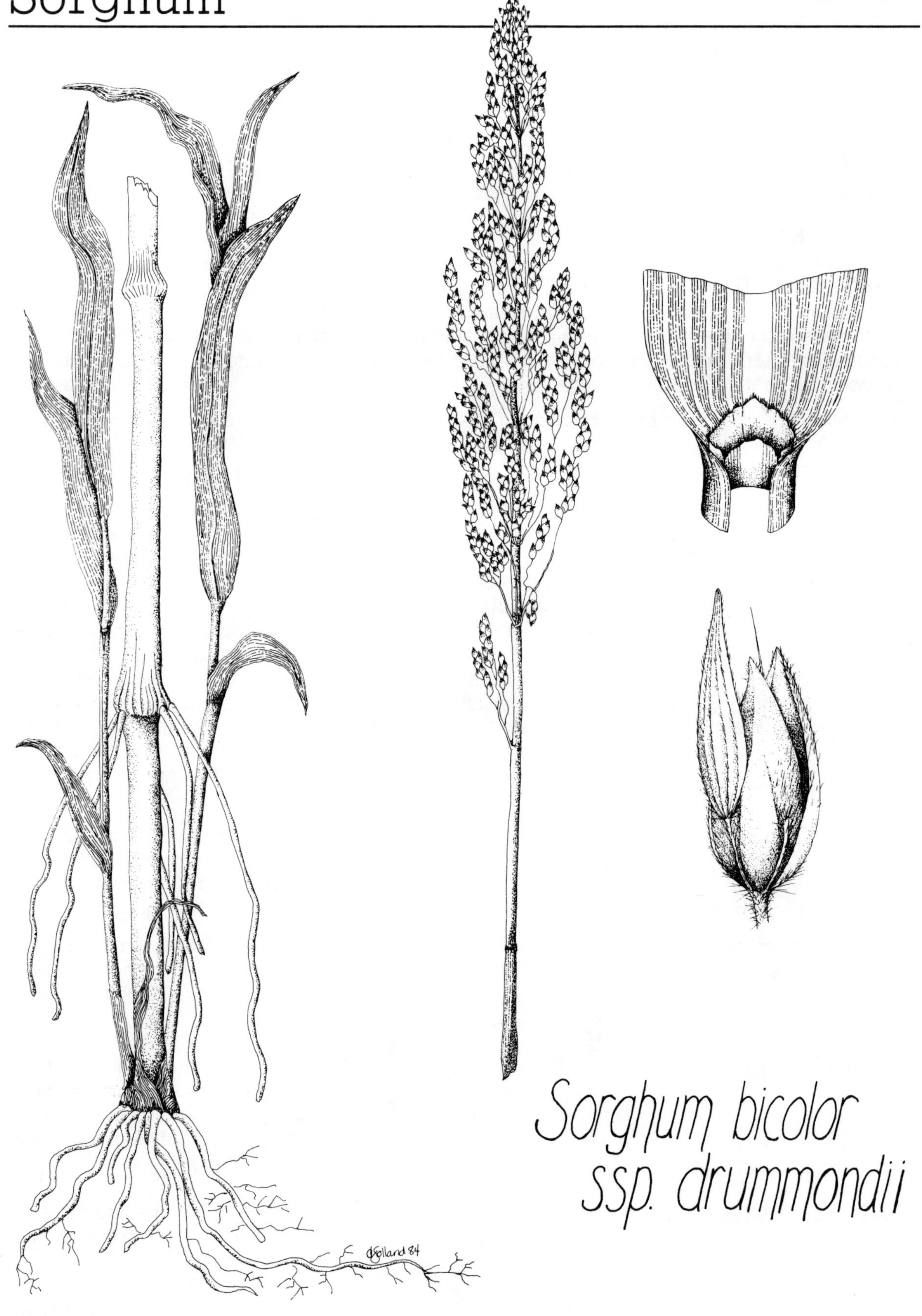
Sorghum bicolor
ssp. drummondii

# GENUS: *Sorghum*

*Sorghum bicolor* (L.) Moench ssp. *bicolor* — Cultivated sorghum
*Sorghum bicolor* (L.) Moench ssp. *drummondii* (Steud.) de Wet — Shattercane
*Sorghum halepense* (L.) Pers. — Johnsongrass

---

**FAMILY:** Gramineae (Poaceae) — the Grass Family (see *Lolium*)

**PHENOLOGY:** Depending on the taxon, environmental conditions and other factors, plants flower from July to September.

**DISTRIBUTION:** Annual sorghum is cultivated widely in the United States. The domesticated types include broomcorn, sudan grass, grain sorghum, forage sorghum, and saccharin sorghum. Johnsongrass, a perennial introduced into the U.S. some time in the early 19th century, has spread as a weed of waste places, railroad yards, highway margins, and cultivated fields. In Pennsylvania it is more commonly encountered in the southeastern quarter of the state. Shattercane is an annual plant that results from crosses between cultivated sorghum and johnsongrass or as a spontaneous appearance of "wild" (ancestral) genes in cultivated sorghum through genetic recombination. Shattercane is more commonly encountered in crop fields.

**PLANT CHARACTERISTICS:** *Sorghum* is a highly variable, diverse group of taxa too complicated to detail here. The **spikelets** are in pairs, numerous, and compressed, forming a large branching panicle, one spikelet of the pair sessile and perfect, the other pedicelled; **glumes:** about equal, hard; **lemmas:** thin, often awned. *Sorghum halepense* is perennial with narrow (4 cm) leaf blades. Members of the genus can attain heights of 10 feet or more.

**POISONOUS PARTS:** Green, aerial portions, especially leaves and stems, (canes) are toxic.

**SYMPTOMS:** Cyanide poisoning to livestock may result from consumption of plants. Mucuous membranes of eyes and mouth may appear congested. Ingesta examined immediately has a characteristic benzaldehyde odor, resulting from the production of benzaldehyde from the breakdown of the aglycone of certain cyanogenic glycosides. Respiration may be stimulated, rapidly altering to dyspnea, excitation, gasping, staggering, paralysis, prostration, convulsions, coma, and death.

Nitrogen poisoning can result from toxic levels of nitrates found in the plants. In ruminant digestion, nitrates are converted to nitrites, which are about ten times more toxic. They are the more immediate cause of poisoning. Symptoms of nitrite toxicosis include cyanosis, severe dyspnea, trembling, and weakness with a chocolate brown discoloration of the blood.

**Postmortem: gross and histological lesions:** bright red blood with congestion of internal organs; serious surface hemorrhage; respiratory passage edema may be present. Horses have been reported to develop chronic cystitis and ataxia and urinary bladder fibrosis in prolonged cases; epithelium ulcerations and abcesses may occur in the wall.

**POISONOUS PRINCIPLES:** Dhurrin, a cyanogenic glycoside, is present in some members of the genus *Sorghum.* Hydrolysis of this compound yields hydrocyanic acid. Forage sorghum also may accumulate levels of nitrates that can cause poisoning. Sheep developed hypersensitivity to light (photosensitization) due to a putatively photodynamic pigment in some species of *Sorghum.* For a complete characterization of photosensitization see *Hypericum* and *Heracleum.*

**CONFUSED TAXA:** As seedlings, virtually all species of *Sorghum* resemble one another and frequently are confused with young corn plants. Annual sorghum is either a cultivated plant or the weed shattercane.

*(Continued)*

# Sorghum

**SPECIES OF ANIMALS AFFECTED:** Cattle mortality was extensive in some regions of the United States in years prior to the development of sorghum strains low in glycosides. Other livestock could be affected by the known sorghum toxins.

**TREATMENT:** (11a)(b); (25)

**OF INTEREST:** Some environmental factors that increase cyanogenic potential include high nitrogen, low phosphorus in soils, drought, and age of plants (young growth having highest potential). Many years of selective breeding have resulted in hybrids having low genetic potential for the development of hydrogen cyanide. Since *S. halepense* is a wild weed, it is to be considered with more suspicion than cultivated varieties. *Sorghum halepense* is a Pennsylvania Legislated Noxious Weed (Act 1982-74). Seeds of it (and those from any crosses) are Restricted Noxious Weed Seeds and must be listed on the tag or label of agricultural seeds for sale in the Commonwealth.

# Tanacetum

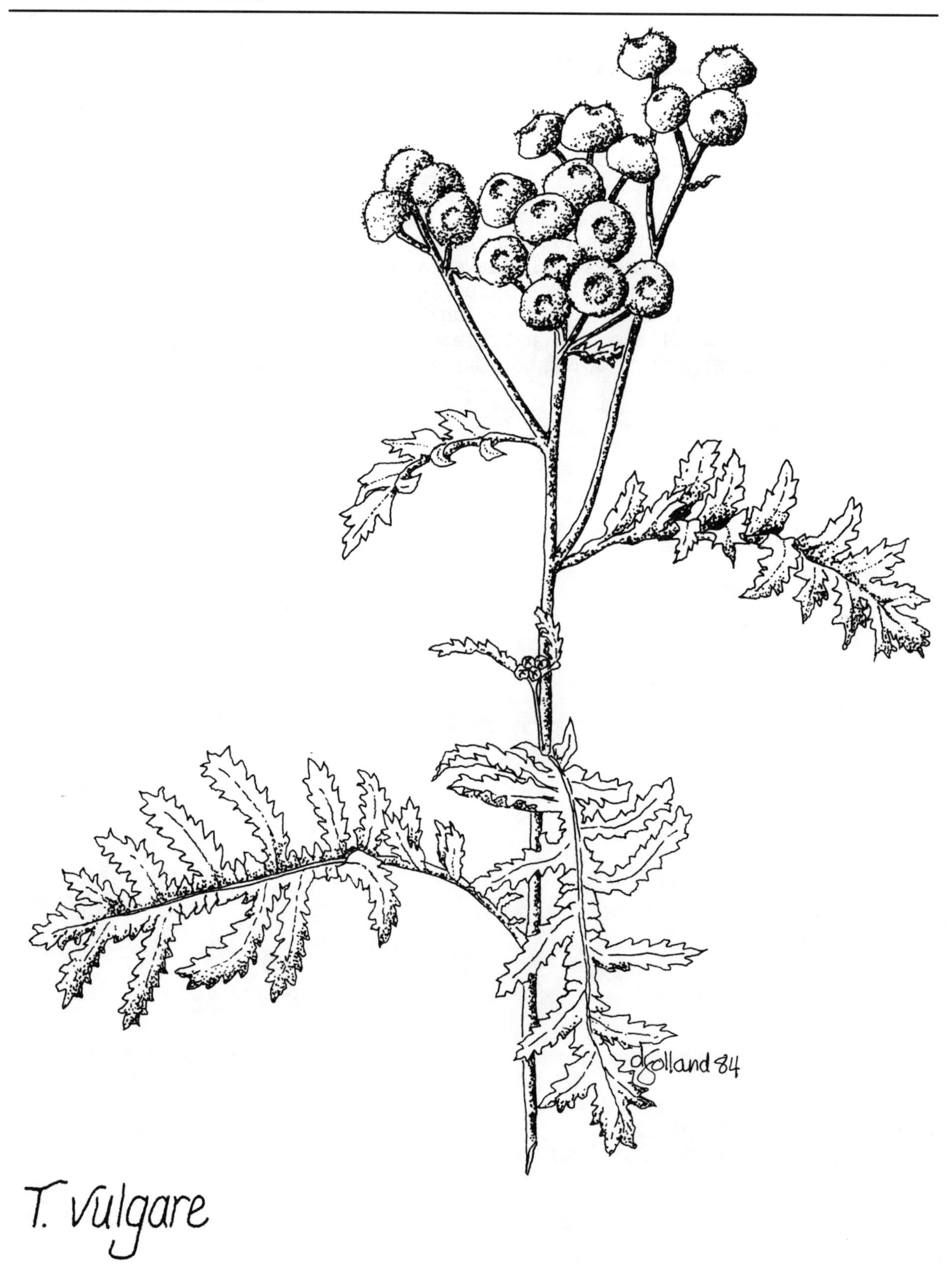

# GENUS: *Tanacetum*

*Tanacetum vulgare* L. - Common tansy

**FAMILY:** Compositae (Asteraceae) — the Daisy Family (see *Arctium*)

**PHENOLOGY:** Tansy is a late-season flowering plant, commonly blooming August through October.

**DISTRIBUTION:** This Old World perennial is found along roadsides, in fields and waste places, and is cultivated in herb and medicinal gardens.

**PLANT CHARACTERISTICS:** *Tanacetum vulgare* can be identified by its coarse, aromatic foliage arising from a stout **rhizome.** The **leaves** are numerous, 1-2 dm long, nearly half as wide, sessile, punctate, pinnatifid, with an evidently winged rachis; **leaflets:** toothed or incised; **flower heads:** many, corymbose, 20-200 disk flowers per head; **disks:** 5-10 mm wide, golden-yellow, 5-toothed; **pappus:** minute, a 5-lobed crown.

**POISONOUS PARTS:** The herbage (leaves and stems) and flowers contain the toxin.

**SYMPTOMS:** Severe gastroenteritis, rapid and weak pulse, violent spasms, convulsions, and death have resulted from overdose of tansy.

**POISONOUS PRINCIPLES:** The source of poisoning is an oil, tanacetin.

**CONFUSED TAXA:** Tansy resembles the pineapple weed (*Matricaria matricarioides* (Less.) Porter), which is an annual, glabrous, pineapple-scented plant with 4-toothed corolla disks. Tansy also resembles costmary (*Chrysanthemum Balsamita* L.), which has simple silvery-strigose leaves.

**SPECIES OF ANIMALS AFFECTED:** The pungent, strong smell of the herbage usually prevents animals from consuming this plant in quantity. Human life has been lost after abuse of medicinal extracts from tansy. Oil of tansy is used as a cure for nervousness, to induce abortion, to foster menstruation, or to kill worms (antihelminthic) in home remedies. Teas made from the herbage can be lethal.

**TREATMENT:** (11a)(b); (26)

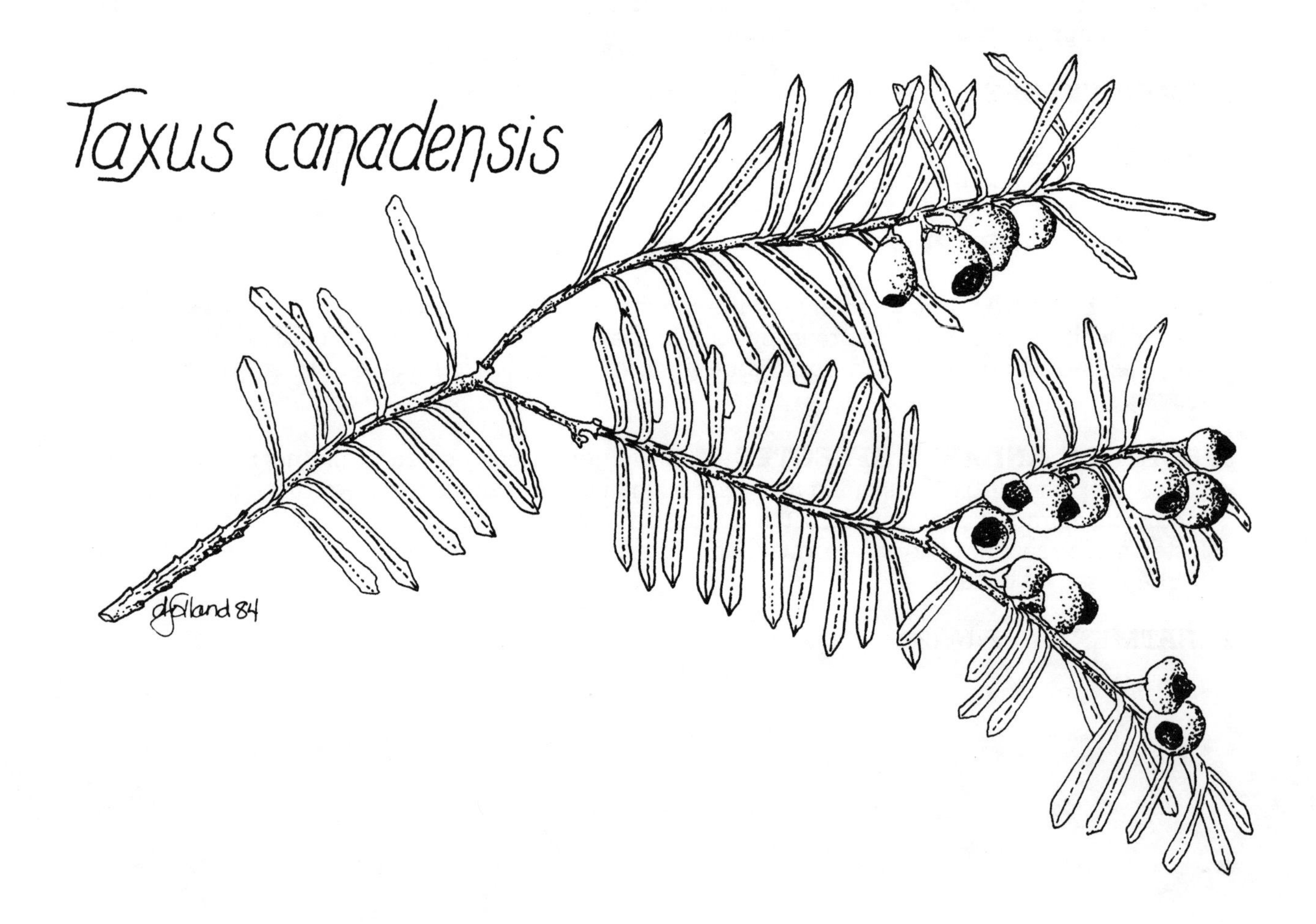
Taxus canadensis

# GENUS: *Taxus*

*Taxus baccata* L. — English Yew
*Taxus canadensis* Marsh — American yew

---

**FAMILY:** Taxaceae — the Yew Family

This group of evergreen shrubs and trees has **leaves:** needlelike, linear or scalelike, often appearing to be 2- ranked; **plants:** gymnosperms, but not producing female cones; **seeds:** surrounded by a hard coat and partly or completely surrounded by a fleshy **aril.**

**PHENOLOGY:** Inconspicuous flowers are produced early in the growing season; the conspicuous, scarlet fruits, late in the season.

**DISTRIBUTION:** *Taxus baccata* is cultivated and planted as an ornamental; more than 40 varieties and forms have been named. *Taxus canadensis* is a northern taxon found in coniferous forests, rich woods, thickets, and bogs. At least four cultivars are used horticulturally.

**PLANT CHARACTERISTICS:** The yews are popular shrubbery grown around the home for landscape value. They can be distinguished by long, slender, alternate dark, glossy green, flat **needles** and by the bright scarlet-red, fleshy cup covering the **fruits.**

**POISONOUS PARTS:** The entire plant contains poisonous alkaloids.

**SYMPTOMS:** Gastric distress, diarrhea, vomiting, tremors, dyspnea, dilated pupils, respiratory difficulties, weakness, fatigue, collapse, coma, convulsions, bradycardia, circulatory failure, and death are the result of ingestion. The toxins are rapidly absorbed by the intestines. Death is sometimes so rapid that few well-developed symptoms appear; survival after poisoning is rare. In Europe, yew is considered the most poisonous tree or shrub.

**POISONOUS PRINCIPLES:** The alkaloid mixture taxine is responsible for poisonings. Taxine I (major alkaloid) and taxine II (minor alkaloid) have been isolated from yews.

**CONFUSED TAXA:** No other woody, evergreen shrub or tree produces the characteristic bright-red, fleshy fruit with an opening at the terminal end.

**SPECIES OF ANIMALS AFFECTED:** Human fatalities due to cardiac and respiratory failure are known. Death in domestic animals is not uncommon. A fatal mistake can be made by placing branches in animal enclosures. Deer will also succomb to ingestion of yews.

**TREATMENT:** (11a)(b); (26); (5- definitely beneficial)

**OF INTEREST:** The red, fleshy aril is sweet and edible, at least in small quantities. *Taxus* has been used medicinally in the past. *Taxus cuspidata* Siebold & Zucc., a species related to those described above, has shown experimental hypoglycemic activity. The cancer chemotherapeutic drug taxol has been extracted from *T. brevifolia* Nutt. Toxicity may vary according to seasonal or geographic factors. All species in the genus should be considered potentially poisonous.

# Urtica

# GENUS: *Urtica*

*Urtica dioica* L. - Stinging nettle

**FAMILY:** Urticaceae — the Nettle Family

In the Commonwealth this family is represented by five genera, including *Urtica* and *Laportea*. Members produce inconspicuous **flowers** that lack petals. The **calyx** is (3-)5-lobed; **stamens:** as many as and opposite to the lobes; **ovary:** 1, superior; **style:** 1; **flowers:** usually unisexual or rarely perfect.

**PHENOLOGY:** Stinging nettles flower June through September.

**DISTRIBUTION:** Moist waste places, roadsides, and rich woods are home to the nettles.

**PLANT CHARACTERISTICS:** Nettles are erect, usually unbranched perennial plants with **leaves:** opposite, serrate, 5-15 cm; **stipules:** lanceolate to linear, 5-15 mm; **inflorescences:** many flowered and branched.

**POISONOUS PARTS:** All parts of the plant can bear stinging hairs. The hairs are fluid-filled, hollow fibers with basal swellings containing the irritant.

**SYMPTOMS:** Contact dermatitis results in an intense burning, itching, or stinging of the skin.

**POISONOUS PRINCIPLES:** Although a combination of histamine, acetylcholine, and 5-hydroxytryptamine has been considered the toxic agent, this recently has been challenged.

**CONFUSED TAXA:** Several varieties of stinging nettle occur here. The well-established European weed, *U. dioica* var. *dioica,* has densely hairy stems and deeply toothed leaves, in contrast to *U. dioica* var. *procera.* Another species in the genus, *U. urens* L., is a stinging weed from Europe becoming established in our range. The genus *Laportea* (wood-nettle) is represented by *L. canadensis* (L.) Wedd., a commonly encountered stinging plant with alternate leaf arrangement.

**SPECIES OF ANIMALS AFFECTED:** It is most troublesome to humans.

**TREATMENT:** (23); (26); in extreme cases (4)

**OF INTEREST:** *Urtica dioica* contains substances with known diuretic properties when consumed. Herb extracts contain volatile oils that are used in cosmetic, botanical hair rinses. Roots and seeds have been an ingredient in hair growth remedies and restorative treatments for hair. Home remedies for relieving the pain caused by stinging nettle include the application of juice from dock *(Rumex),* which often grows in the same habitat, and from onion, leek, or plantain. In spring, short, tender, young plants can be gathered for eating. The plant is reputedly high in protein and has a delicate flavor. Boiling water is said to quell the stinging hairs.

# GENUS: *Veratrum*

*Veratrum viride* Ait. — False hellebore

---

**FAMILY:** Liliaceae — The Lily Family (see *Amianthium*)

**PHENOLOGY:** False hellebore flowers June to July.

**DISTRIBUTION:** This plant grows in swamps, low wet places, meadows, pastures, and open woods.

**PLANT CHARACTERISTICS:** *Veratrum viride* is a coarse, tall, unbranched **herb,** 3 to 6 feet, perennial from a short rhizome; **leaves:** large (appearing pleated), alternate in 3-ranks, broad, the bases sheathing the stems; **panicle:** terminal, composed of greenish-yellow to purple, hairy flowers, about 1.5 cm across; **tepals:** 6, narrowed at base, not glandular; **stamens:** 6, filaments free from the perianth; **ovary:** tri-lobed, each lobe terminating in a short style; **fruit:** an ovoid capsule, surrounded by the withered perianth; **seeds:** large, flat, the embryo small and surrounded by a broad wing. See *Amianthium* for illustration.

**POISONOUS PARTS:** All parts are poisonous, especially the young, succulent growth in spring.

**SYMPTOMS:** Species vary in physiologically active principles, yet symptoms of acute poisoning are constant: salivation, vomiting, diarrhea, stomach pains, prostration, depressed heart action, general paralysis, spasms, and dyspnea. Death may result. In addition, hallucinations, headache, and a burning sensation of mouth and throat have been reported. A species of *Veratrum* from western United States is known to cause congenital malformation in lambs, including cyclopia (single median eye) and cranial and lower jaw abnormalities. Ewe embryos in the primitive streak stage (12th and 14th day of gestation) develop deformities; fetal pituitary may be absent. **Postmortem: gross and histological lesions:** not reported in acute toxicity.

**POISONOUS PRINCIPLES:** The numerous known alkaloids exist as glyco - or ester alkaloids and include jervine, pseudojervine, rubijervine, cevadine, germitrine, germidine, veratralbine, and veratroidine. Plants also may contain cardiac glycosides.

**CONFUSED TAXA:** The plant is readily recognized, although the name can be confused with true hellebore (see *Helleborus*) in the Ranunculaceae.

**SPECIES OF ANIMALS AFFECTED:** Humans and all classes of livestock, especially cattle, sheep, and fowl.

**TREATMENT:** (11a)(b); (1); (5); (12)

**OF INTEREST:** For two centuries, until the 1900's *Veratrum viride* was used widely in medicine and as an insecticide. It is now the source of a class of antihypertensive agents that effect the afferent side of the sympathetic nervous system. Rootstock extracts used in homeopathy should be used cautiously since this plant contains teratogens, substances causing fetal deformations. Most accounts of false hellebore poisoning have from been medicinal misuse. It is used in veterinary medicine as a circulatory depressant, stomachic, emetic, and parasiticide.

# GENUS: *Wisteria*

*Wisteria* spp. — Wisteria

**FAMILY:** Fabaceae (Leguminosae) — the Bean (Legume) Family (see *Crotalaria*)

**PHENOLOGY:** Most wisteria plants flower from April through May; Japanese wisteria blossoms through June.

**DISTRIBUTION:** Wisteria is commonly used as an ornamental twining shrub near homes and patios or as a specimen plant on lawns. Several species are native to North America, occurring at borders of wooded swamps and banks of streams onto rich woods. Two species have been introduced from eastern Asia.

**PLANT CHARACTERISTICS:** Wisteria flowers have the 2 upper **calyx**-lobes short or completely fused, with the lowest one often elongate. The **standard** is reflexed, with two basal hardened thickenings; **flowers:** purple (blue) or white in grapelike clusters at the end of short branches, appearing before the leaves are fully expanded; **seed pods:** flattened, woody, often covered with velvety hairs; **seeds:** resembling large lima beans; **plants:** woody, twining shrubs or vines with odd-pinnate leaves.

**POISONOUS PARTS:** The seeds and seed pods are toxic.

**SYMPTOMS:** Toxicosis involves digestive disturbances, stomach and intestinal irritation, repeated vomiting, diarrhea, abdominal pain, and collapse.

**POISONOUS PRINCIPLES:** The toxins are unidentified.

**CONFUSED TAXA:** Wisterias are woody plants with compound leaves. In *W. frutescens* (L.) Poir and *W. macrostachya* Nutt. the leaflets number 9-15. *Wisteria sinensis* (Sims) Sweet (Chinese wisteria) usually has 11 leaflets. *Wisteria floribunda* (Willd.) DC (Japanese wisteria) has 15-19 leaflets. No woody vines other than members of this genus produce the combination of grapelike clusters of flowers and velvety seed pods.

**SPECIES OF ANIMALS AFFECTED:** Children are poisoned from eating the lima beanlike fruit. One or two seeds can cause serious illness.

**TREATMENT:** (11a)(b); (26)

# Xanthium

*X. strumarium*

# GENUS: *Xanthium*

*Xanthium* spp. — Cocklebur

**FAMILY:** Compositae (Asteraceae) — the Daisy Family (see *Arctium*)

**PHENOLOGY:** Cocklebur flowers July through October.

**DISTRIBUTION:** It is found in fields, waste places, flood plains, and on lake and sea beaches.

**PLANT CHARACTERISTICS:** These include **heads:** small, unisexual; **staminate heads:** uppermost, many-flowered; **involucre:** of separate bracts in 1-3 series; **receptacle:** cylindric, chaffy; **filaments:** monadelphous; **pistillate heads:** completely enclosing the two flowers, forming a conspicuous 2-chambered bur with hooked prickles; **pistillate flowers:** corolla absent; styles exsert from the involucre; **achenes:** thick, solitary in chambers of the bur; **pappus:** absent; **plants:** annuals, coarse, weedy; **leaves:** coarse, alternate.

**POISONOUS PARTS:** Cocklebur poisoning is always associated with ingestion of seedlings. Seeds and seedlings are toxic and remain toxic when dry. Seeds are toxic at a rate of 0.3% of the animal's body weight; seedlings, at a rate of 1.5%.

**SYMPTOMS:** Anorexia, depression, nausea, vomition, weakened heartbeat, muscular weakness, dyspnea, and prostration are symptoms of poisoning. Spasmodic running motions or convulsions are observed when animals are prostrate. Also, abdominal pain is present in pigs. Fowl, which may be poisoned by the seeds, show profound depression. Symptoms may appear within 2 days; death occurs within a few hours to 3 days after a toxic dose. **Postmortem: gross lesions:** gastroenteric irritation and congestion; serofibrinous ascites, gallbladder wall edema; accentuation of liver lobes; **histological lesions:** acute, diffuse, centrilobular and paracentral coagulative necrosis of the liver.

**POISONOUS PRINCIPLES:** The agent of toxicosis, hydroquinone, is used commercially in photographic processes. Ingestion of 1 g has caused tinnitus, nausea, vomiting, sense of suffocation, shortness of breath, cyanosis, convulsions, delirium, and collapse. Death has followed ingestion of 5 g.

**CONFUSED TAXA:** Burdock *(Arctium)* may be confused with cocklebur *(Xanthium)*; however, the involucral bracts in the former are distinctly hooked at the tip, a feature not found in the latter (see *Arctium*).

**SPECIES OF ANIMALS AFFECTED:** All species of livestock are susceptible.

**TREATMENT:** (11a)(b); (26)

# GLOSSARY

## A

**Abaxial** — away from the axis (dorsal)
**Abortifacient** — an agent causing loss of a fetus
**Abortion** — imperfect development or loss of a part
**Acaulescent** — stemless; floral stalk may be present
**Accrescent** — enlarging after flowering
**Achene** — indehiscent, 1-seeded fruit
**Actinomorphic** — radially symmetrical flower; can be bisected, on many planes, into equal halves; compare with zygomorphic
**Acuminate** — tapering at the end to a gradual point
**Acute** — terminating in a sharp angle
**Adaxial** — toward the axis (ventral)
**Adnate** — different organs fused, e.g. stamens to petals
**Adventive** — a foreign plant not widely established
**Alternate** — borne at regular intervals at different levels (not opposite to each other)
**Androecium** — a collective term for stamens
**Annual** — germinating, growing, flowering, seeding in one season
**Anorexia** — loss of appetite
**Anther** — the part of the male organ (stamen) which produces pollen
**Anthesis** — the time of flowering
**Anuria** — absence of urination
**Apetalous** — without petals
**Aril** — fleshy appendage surrounding a seed
**Ascites** — accumulation of fluid in the abdomen
**Asphyxia** — lack of oxygen in the body
**Ataxia** — an inability to coordinate voluntary muscle movements
**Atelectasis** — collapse of a portion of a lung
**Atonic** — without normal muscle tone
**Arthrogryposis** — retention of a joint in a fixed position
**Awn** — a bristle

## B

**Banner** — upper petal of a legume flower
**Berry** — a pulpy fruit with immersed seeds
**Biennial** — of two seasons duration from germination to seed-set
**Bifid** — divided into two parts (clefts)
**Bilateral** — arranged on opposite sides; see zygomorphic
**Bipinnate** — doubly or twice pinnate
**Bisexual** — having both male (stamens) and female (pistil) organs; hermaphroditic
**Blade** — the expanded portion of a leaf
**Bract** — a reduced leaf beneath a flower or inflorescence
**Bradycardia** — slow heart action
**Bulb** — a subterranean leaf-bud with fleshy scales

## C

**Caducous** — falling off very early
**Calyx** — the outer envelope of the flower; collective term for the sepals
**Carpel** — a simple pistil or one element of a pistil
**Catarrh** — inflammation of a mucous membrane usually affecting the nose and air passage
**Ciliate** — the margin fringed with hairs
**Circumscissile** — splitting in a horizontal, circular line
**Cirrhosis** — interstitial inflammation of any tissue or organ
**Claw** — the narrowed base of some petals
**Coma** — a tuft of soft silky hairs
**Compound** — composed of 2 or more similar structures; e.g. a compound leaf is divided into separate leaflets
**Congestion** — a condition of being clogged or causing an excessive fullness of the blood vessels
**Connate** — fusion of like structures, e.g. staminal filaments in the legume flower
**Contact dermititis** — causing inflammation of the skin upon contact
**Cordate** — heart-shaped
**Corm** — a solid subterranean bulb-like structure
**Corolla** — the collective term for petals
**Corona** — a crown or inner petal-like structure
**Corymb** — a flat-topped flower-cluster; floral stalks not originating at the same point
**Crenate** — having rounded teeth
**Culm** — the stem of grasses
**Cyanosis** — turning blue or purple due to a lack of oxygen

**Cyme** — a flower-cluster; usually broad and flattish, with its central flower blooming first
**Cystitis** — inflammation of the urinary bladder

## D

**Deciduous** — not evergreen; dropping
**Decompound** — more than once compound or divided
**Dentate** — toothed
**Dioecious** — unisexual flowers with the two kinds of flowers on separate plants; compare with monoecious
**Disk-flower** — the tubular flowers in a composite (daisy) head; compare with ray-flower
**Dissected** — cut or divided into narrow segments
**Drupe** — a fleshy fruit with a hard, stony center
**Dyspnea** — shortness of breath; difficult or labored respiration

## E

**Ecchymoses** — discoloration of the skin or soft tissue caused by invasion of blood
**Edema** — excessive accumulation of body fluid in tissues or cavities
**Elliptical** — rounded about equally at both ends
**Emphysema** — air-filled expansions of body tissue
**Enteritis** — inflammation of the intestines
**Entire** — lacking teeth or divisions
**Epigynous** — stamens, petals, and sepals fused on top of the ovary
**Escaped** — cultivated, but spreading without cultivation

## F

**Filament** — the portion of the male organ (stamen) which supports the anther
**Floret** — a small flower, often one of a dense cluster

## G

**Gamopetalous** — having the petals united (fused)
**Gamosepalous** — having the sepals united (fused)
**Gastroenteritis** — inflammation of the stomach and intestines
**Glabrous** — without hairs; smooth
**Gynoecium** — a collective term for the pistils of the flower

## H

**Haustoria** — rootlike structures modified for nutrient absorption in parasitic plants
**Head** — a dense cluster of sessile flowers
**Hematuria** — blood or blood cells in the urine
**Hemoglobinuria** — having hemoglobin (a blood pigment) in the urine
**Hemolytic anemia** — anemia resulting from the excessive destruction of red blood cells
**Hemorrhage** — a copious discharge or loss of blood
**Hydroperitoneum** — fluid in the peritoneal (abdominal) cavity
**Hydrothorax** — fluid in the pleural (lung) cavity
**Hypanthium** — an enlargement of the receptacle under the sepals or the fused basal portion of floral parts
**Hyperemia** — a condition resulting in distention of the blood vessels caused by increased blood in a body part
**Hypogynous** — sepals, petals, and stamens fused under the ovary

## I

**i.m.** — intramuscular; administered into the muscle tissue; also written IM
**i.p.** — intraperitoneal; administered into the body cavity; also written IP
**i.v.** — intravenous; administered into the veins; also written IV
**Icterus** — jaundice; a yellowish coloration of skin, tissues, and body fluids
**Imperfect** — have only one functional kind of sex organ, e.g. either male (stamens) or female (pistils)
**Inferior** — see epigynous
**Inflorescence** — the flowering shoot bearing more than one flower; a floral cluster
**Introduced** — a plant of foreign origin brought to a new region for cultivation, or arriving accidentally
**Irregular** — showing inequality in size, form, or union

## K

**Ketosis** — an accumulation of metabolic poisons

# L

**Labiate** — lipped
**Lacrymal** — relating to the secretion of tears
**Lanceolate** — several times longer than wide; broadest toward the base
**$LD_{50}$** — the lethal dose in which 50% of the test animals die from an experimental treatment
**Leaflet** — a single leaf-like division of a compound leaf
**Lemma** — the lower of the two bracts inclosing the flower in grasses
**Ligule** — see ray-flower
**Limb** — the expanded portion of a petal or leaf
**Linear** — long and narrow with parallel sides
**Lip** — the upper and lower divisions of a bilabiate corolla
**Lobed** — divided into segments

# M

**-merous** — used in combination with numbers to describe the multiples of each organ; e.g. 4-merous implies 4 sepals, 4 petals, 4 or 8 or 12 etc. stamens, 4 carpels
**Methemoglobinemia** — a condition in which the blood is incapable of carrying oxygen
**Monodelphous** — staminal filaments united into a tube
**Monoecious** — species with unisexual flowers on the same plants; compare with dioecious
**Mucosa** — a mucous membrane; tissue which lines the body cavity and passages
**Mydriasis** — prolonged or excessive dilatation of the pupil of the eye

# N

**Naturalized** — a foreign plant becoming thoroughly established
**Necrosis** — localized death of living tissue
**Node** — the place on a stem which bears leaves

# O

**Oblanceolate** — lanceolate with the broadest part at the apex
**Oblong** — two (or three) times longer than wide with nearly parallel sides
**Obtuse** — blunt (rounded) at the end
**Ovary** — the female reproductive organ
**Ovate** — egg shaped

# P

**Palea** — in grasses, the upper bract that together with the lemma encloses the flower
**Palmate** — radiately lobed or divided; divisions from a single point
**Panicle** — a branched, pedicellate flower-cluster
**Pappus** — the modified calyx of a composite flower, below the corolla and above the achene
**Pedicel** — the flower-stalk of a single flower
**Peduncle** — a primary flower-stalk supporting either a cluster or a solitary flower
**Pendulous** — hanging down
**Percutaneous** — absorption through the skin
**Perennial** — living for more than two seasons
**Perfect** — having both functional male (stamens) and female (pistils) organs in the same flower
**Perianth** — the floral envelope consisting of the calyx and corolla
**Petal** — the division of the corolla
**Petaloid** — resembling a petal
**Petechiae** — small, rounded hemorrhages on the surface of the skin, mucous membrane, serous membrane, or on a cross-sectional surface of an organ
**Petiolate** — having a petiole
**Petiole** — the support (stalk) of a leaf
**Phenology** — the study of periodic biological phenomena, such as flowering time, especially as it relates to climate
**Pilose** — having soft hairs
**Pinnatifid** — pinnately cleft
**Pip** — an underground stem resembling a root
**Pistil** — the seed producing organ of the flower composed of ovary, stigma and style
**Plicate** — folded, usually lengthwise, into plaits
**Plumose** — having hairs like the plume of a feather
**Polypetalous** — having separate petals
**Pubescent** — covered with hairs
**Punctate** — dotted with pits, translucent internal glands or colored pits
**Purgative** — a substance causing evacuation of the bowels

# R

**Raceme** — a simple flower-cluster of pedicelled flowers upon a common, more or less elongated axis
**Rachis** — the axis of a flower-cluster or of a compound leaf

**Ray** — the strap-like marginal flower of the Compositae; compare with disk-flower
**Receptacle** — the region at the end of the flower-stalk to which the flower is attached
**Regular** — uniform in shape or structure; see actinomorphic
**Reniform** — kidney-shaped
**Retrorse** — directed downward
**Rhizome** — any prostrate or subterranean stem, rooting at the nodes, resembling a root
**Rosette** — a cluster of leaves in a circular form, often at ground-level
**Rotate** — flat and circular in outline
**Rotund** — rounded in outline

# S

**Scape** — a naked flowering stalk arising from the ground
**Scoliosis** — lateral curvature of the spine
**Segment** — one of the parts of a leaf that is cleft or divided
**Sepal** — a division of a calyx
**Sepaloid** — resembling a sepal
**Septate** — divided by a septum (partition)
**Serrate** — having sharp teeth pointing forward
**Sessile** — without a stalk
**Sheath** — the tubular envelope of a leaf, especially the lower part of a grass leaf
**Shrub** — a woody perennial, smaller than a tree, usually with several stems at ground level
**Silique** — an elongate (fruit) capsule divided into two valves in the mustard family; when spherical or shortened it is called a silicle
**Spadix** — a spike with a fleshy axis in the aroid family, bearing flowers
**Spathe** — a large leaf-like bract enclosing a spadix
**Spike** — a simple flower-cluster with sessile flowers
**Spikelet** — a small spike
**Spur** — a hollow sac-like or tubular extension of the sepals or petals or filaments, usually nectariferous
**Stamen** — the pollen-bearing organ of a flower composed of a stalk (filament) and pollen producing sacs (anthers)
**Standard** — the upper dilated petal of a legume flower
**Stigma** — the part of a pistil or style that receives the pollen for effective fertilization
**Stipulate** — having stipules
**Stipule** — a leaf-like appendage at the base of a petiole or leaf
**Stolon** — a horizontal runner, or basal branch rooting at the nodes
**Style** — the attenuated portion of the pistil connecting the stigma to the ovary
**Subsessile** — minutely stalked
**Superior ovary** — see hypogynous
**Symmetrical** — uniform number of floral parts
**Sympetalous** — see gamopetalous

# T

**Tachycardia** — rapid heart action
**Taxa** — plural of taxon
**Taxon** — any unit or classification, e.g. genus, species, etc.
**Tepal** — a collective term for the floral envelope when petals and sepals are not readily differentiated
**Tetanic** — muscle spasms or prolonged contraction of a muscle
**Tinnitus** — a sensation of noise, as a ringing sound
**Torticollis** — deformity of the neck resulting in abnormal position and limited movements of the head
**Toxicosis** — a pathological condition caused by the action of a poison or toxin
**Tree** — a perennial woody plant with an evident trunk
**Tuber** — a subterranean branch, resembling a short, thick root, usually having numerous buds called "eyes"

# U

**Umbel** — a flower-cluster in which the peduncles or pedicels arise from the same point
**Uremia** — accumulation of urine constituents in the blood
**Umbellet** — a secondary umbel

# W

**Whorl** — an arrangement of leaves in a circle around the stem

# Z

**Zygomorphic** — capable of division in only one plane of symmetry; compare with actinomorphic

# APPENDIX I — TREATMENTS

## GENERAL FIRST AID

When a child eats any non-food plant material, contact a physician or a poison control center whether symptoms are present or not. If a poisonous plant has been eaten and if medical help is not readily available, the U.S. Department of Health and Human Services, Division of Poison Control suggests the following: vomiting should be induced (unless the victim is already vomiting, is unconscious or convulsing) by giving syrup of Ipecac with a glass of water (1 tablespoonful or 15 ml or 1/2 ounce of Ipecac for children, double for adults). To prevent the vomitus from being inhaled (aspirated) into the lungs the victim should be made to walk about, or should be held in a head-down "spanking position". After vomiting has ceased, about 1 ounce (child; 3 oz. adult) of activated charcoal and water should be given orally. Because activated charcoal will not dissolve in water, it should be swirled around in the glass and drunk quickly. Many pediatricians recommend that syrup of Ipecac and activated charcoal be kept in the home medicine chest.

If the child must be brought to a physician or emergency room, a sample of the plant, with flowers and seeds if possible, should be brought along for positive identification.

## TREATMENT CATEGORIES FOR THE HEALTH CARE ATTENDANT

(1) Charcoal
(2) Anti-convulsents (e.g. parenteral short-acting barbiturates)
(3) Antidiarrheal agents
(4) Antihistamines
(5) Atropine
(6) Artificial respiration and oxygen
(7) Calcium gluconate, parenteral fluids
(8) Control pain with Demerol
(9) Demulcents
(10) Epinephrine
(11) a) Gastric lavage, or b) Emesis
(12) Hypotensive Drugs
(13) Keep urine alkaline
(14) Mineral or castor oil
(15) Narcotic antagonistic drugs
(16) Oral doses of strong tea or tannin
(17) Paraldehyde (2-10cc) IM
(18) Pilocarpine or physostigmine for dry mouth and visual disturbances
(19) Potassium, procainamide, quinidine sulfate, disodium salt of edetate ($Na_2EDTA$)
(20) Saline cathartic
(21) Sedation
(22) Shock therapy
(23) Steroidal creams; topological ointments for inflammation of skin, blisters
(24) Stimulants including strong coffee
(25) Treat for cyanide poisoning
(26) Treat symptoms as they appear, supportive therapy
(27) Maintain blood pressure

# APPENDIX II — POISONOUS STATUS OF BERRIES AND FRUITS

## A. POISONOUS

*Actaea* spp. (Baneberry, dolls-eyes) — POISONOUS
*Aralia* spp. (Sarsaparilla) — POISONOUS RAW; see REPORTED EDIBLE entry below
*Caulophyllum thalictroides* (Blue Cohosh) — POISONOUS RAW; see REPORTED EDIBLE entry below
*Celastrus scandens* (Bittersweet) — POISONOUS
*Convallaria majalis* (Lily-of-the-Valley) — POISONOUS
*Euonymus* spp. (Burning bush, wahoo) — POISONOUS
*Hedera helix* (Ivy) — POISONOUS
*Malus* spp. (Apple, Crabapple) — POISONOUS seeds; fruit edible
*Menispermum canadense* (Moonseed) — POISONOUS
*Parthenocissus quinquefolia* (Virginia creeper) — POISONOUS
*Phoradendron serotinum* (Mistletoe) — POISONOUS
*Phytolacca americana* (Pokeweed) — POISONOUS RAW; see REPORTED EDIBLE entry below
*Podophyllum peltatum* (May apple) — POISONOUS when immature; see REPORTED EDIBLE entry below
*Prunus* spp. (Chokecherry, wild black cherry) — POISONOUS pit; pulp edible
*Rhodotypos scandens* (Jetbead bush) — POISONOUS
*Rhus radicans* (Poison ivy) — POISONOUS
*Rhus Vernix* (Poison sumac) — POISONOUS
*Sambucus* spp. (Elderberry) — POISONOUS roots, stems, leaves; WARNING, uncooked berries may produce nausea; see REPORTED EDIBLE entry below
*Solanum* spp. (Nightshade) — POISONOUS
*Taxus canadensis* (Yew) — POISONOUS seeds; red flesh edible

## B. POTENTIALLY POISONOUS

*Asimina triloba* (Custard apple, pawpaw) — WARNING, some individuals are sensitive to the fruit
*Asparagus officinalis* (Asparagus) — WARNING, red berries may cause toxicosis for some individuals
*Cornus* spp. (Dogwood) — WARNING, may cause toxicosis
*Ilex* spp. (Holly) — WARNING, large quantities (esp. for small children) are dangerous
*Polygonatum* spp. (Solomon's seal) — WARNING, berries contain an anthraquinone causing vomiting/diarrhea
*Smilacina racemosa* (False Solomon's seal) — WARNING, suspected toxic

## C. REPORTED EDIBLE

*Amelanchier* spp. (Service-berry, shadbush) — Raw; cooked
*Aralia* spp. (Sarsaparilla) — Eaten when cooked as jelly
*Arisaema triphyllum* (Jack-in-the-pulpit) — Raw; peppery
*Asimina triloba* (Custard apple, pawpaw) — Raw; pies
*Asparagus officinalis* (Asparagus) — Raw; cooked
*Berberis* spp. (Barberry) — Raw; wine; jelly; pies
*Caulophyllum thalictroides* (Blue cohosh) — Roasted seeds used as a coffee substitute
*Celtis occidentalis* (Hackberry) — Raw; dried
*Cornus* spp. (Dogwood) — Cooked; raw
*Crataegus* spp. (Hawthorn) — Raw; jelly
*Diospyros* spp. (Persimmon) — Raw; pies; pudding

*Duchesnia indica* (Indian strawberry) — Insipid; raw; cooked; jelly
*Fragaria* spp. (Strawberry) — Raw; pies; jelly
*Gaultheria* spp. (Wintergreen) — Raw; cooked
*Ginkgo biloba* (Gingko) — Flesh putrid (fetid); kernel edible
*Juniperus* spp. (Juniper) — Raw; flavoring
*Lonicera* spp. (Honeysuckle) — Insipid; raw; cooked; dried
*Lycium halimifolium* (Matrimony-vine) — Raw; cooked; dried
*Mahonia* spp. (Oregon grape) — Raw; jelly
*Mitchella repens* (Partridge-berry) — Insipid; raw
*Monstera deliciosa* (Monstera) — Raw, if fully ripe
*Morus* spp. (Mulberry) — Raw, if fully ripe; pies; jelly
*Panax quinquefolium* (Ginseng)—Insipid; raw
*Peltandra virginica* (Arrow arum) — Dried; boiled
*Physalis* spp. (Ground-cherry) — Jelly, if ripe
*Phytolacca americana* (Pokeweed) — Edible cooked
*Podophyllum peltatum* (May apple) — Edible if fully ripe
*Pontederia cordata* (Pickerel-weed) — Raw; boiled; dried
*Rhus typhina* (Stag-horn sumac) — Infusion (tea)
*Ribes* spp. (Currant, gooseberry) — Raw; jelly
*Rosa* spp. (Rose) — Raw; jelly; infusion (tea)
*Rubus* spp. (Raspberry) — Raw; jelly; pies; wine
*Sambucus* spp. (Elderberry) — Edible, if fully ripe
*Shepherdia* spp. (Buffalo-berry) — Raw; jelly; drink
*Smilax* spp. (Green briar) — Raw
*Taxus canadensis* (Yew) — Edible aril; see POISONOUS entry above
*Vaccinium* spp. (Blueberry, cranberry) — Raw; jelly; pies; wine
*Viburnum* spp. (Viburnum) — Raw; jelly; pies
*Vitis* spp. (Grape) — Raw; jelly; wine

# REFERENCES

Darbaker, L. K. 1940. *Some poisonous plants of western Pennsylvania.* Mimeographed Circular. University of Pittsburgh, Pittsburgh, PA. 14pp.

Ditmer, Wendell P. 1965. *Poisonous plants of Pennsylvania.* Pennsylvania Department of Agriculture, Bureau of Plant Industry, Harrisburg, PA. 51 pp.

Graham, Edward H. 1935. *Poisonous plants of Pennsylvania.* Carnegie Museum, Pittsburgh, PA. 15 pp.

Gress, E. M. 1935. *Poisonous plants of Pennsylvania.* Bull. No. 531, Pennsylvania Department of Agriculture, Vol. 18 (5): 52 pp.

Hardin, James W. and Jay M. Arena. 1974. *Human poisoning from native and cultivated plants.* Duke University Press, Durham, North Carolina. 194 pp.

Kinghorn, A. Douglas, Edit. 1979. *Toxic plants.* Columbia University Press, New York. 195 pp.

Kingsbury, John M. 1964. *Poisonous plants of the United States and Canada.* Prentice-Hall, Inc., Englewood Cliffs, NJ. 626 pp.

Lewis, Walter H. and M. P. F. Elvin-Lewis. 1977. *Medical botany. Plants affecting man's health.* John Wiley & Sons, New York. 515 pp.

Scimeca, Joseph M. and Frederick W. Oehme. 1985. Postmortem guide to common poisonous plants of livestock. *Vet. Hum. Toxicol.* 27(3): 189-199.

# INDEX TO TAXA

## A



*Page numbers in italics refer to illustrations.*

## B



## C

*Page numbers in italics refer to illustrations.*

# D



# E



# F

## G



## H



*Page numbers in italics refer to illustrations.*



## I



## J



## K



## L

# M



*Page numbers in italics refer to illustrations.*

# N



# O



# P

## Q



## R



## S



## T



*Page numbers in italics refer to illustrations.*

# U



# V



# W



# X



# Y



*Page numbers in italics refer to illustrations.*